Novel Methods for Solving Linear and Nonlinear Integral Equations

Novel Methods for Solving Linear and Nonlinear Integral Equations

Santanu Saha Ray
Prakash Kumar Sahu

CRC Press
Taylor & Francis Group
Boca Raton London New York

CRC Press is an imprint of the
Taylor & Francis Group, an **informa** business

CRC Press
Taylor & Francis Group
6000 Broken Sound Parkway NW, Suite 300
Boca Raton, FL 33487-2742

Printed on acid-free paper
Version Date: 20181003

International Standard Book Number-13: 978-1-138-36274-1 (Hardback)

Library of Congress Cataloging-in-Publication Data

Names: Saha Ray, Santanu, author. | Sahu, Prakash Kumar, author.
Title: Novel methods for solving linear and nonlinear integral equations / Santanu Saha Ray, Prakash Kumar Sahu.
Description: Boca Raton : CRC Press, Taylor and Francis Group, 2019. | Includes bibliographical references and index.
Identifiers: LCCN 2018028398| ISBN 9781138362741 (hardback : alk. paper) | ISBN 9780429431906 (ebook)
Subjects: LCSH: Integral equations--Numerical solutions. | Wavelets (Mathematics)
Classification: LCC QA431 .S165 2019 | DDC 515/.724--dc23
LC record available at https://lccn.loc.gov/2018028398

Visit the Taylor & Francis Web site at
http://www.taylorandfrancis.com

and the CRC Press Web site at
http://www.crcpress.com

To My Parents

Prakash Kumar Sahu

Contents

Preface

Integral equation has been one of the essential tools for various areas of applied mathematics. In this work, we review different numerical methods for solving both linear and nonlinear Fredholm integral equations. A goal is to categorize the selected methods and assess their accuracy and efficiency. We discuss challenges faced by researchers in this field, and we emphasize the importance of interdisciplinary effort for advancing the study on numerical methods for solving integral equations. Integral equations can be viewed as equations which are results of transformation of points in given vector spaces of integrable functions by the use of certain specific integral operators to points in the same space. If, in particular, one is concerned with function spaces spanned by polynomials for which the kernel of the corresponding transforming integral operator is separable being composed of polynomial functions only, then several approximate methods of solution of integral equations can be developed.

This work, especially, deals with the development of different wavelet methods for solving integral and intgro-differential equations. Wavelets theory is a relatively new and emerging area in mathematical research. It has been applied in a wide range of engineering disciplines; particularly, wavelets are very successfully used in signal analysis for waveform representations and segmentations, time frequency analysis, and fast algorithms for easy implementation. Wavelets permit the accurate representation of a variety of functions and operators. Moreover, wavelets establish a connection with fast numerical algorithms. Wavelets can be separated into two distinct types, orthogonal and semi-orthogonal.

The preliminary concept of integral equations and wavelets is first presented in Chapter 1. Classification of integral equations, construction of wavelets and multi-resolution analysis (MRA) have been briefly discussed and provided in this chapter. In Chapter 2, different wavelet methods are constructed and function approximation by these methods with convergence analysis has been presented.

In Chapter 3, linear semi-orthogonal compactly supported B-spline wavelets together with their dual wavelets have been applied to approximate the solutions of Fredholm integral equations (both linear and nonlinear) of the second kind and their systems. Properties of these wavelets are first presented; these properties are then utilized to reduce the computation of integral equations to some algebraic equations. Convergence analysis of the B-spline method has been discussed in this chapter. Again, in Chapter 4, the system of nonlinear Fredholm integral equations has been solved by using hybrid Legendre Block-Pulse functions and the Bernstein collocation method. In Chapter 5, two practical problems arising from chemical phenomena have been modeled as Fredholm- Hammerstein integral equations and solved numerically by different numerical techniques. First, the COSMO-RS model has been solved by the Bernstein collocation method, the Haar wavelet method and the Sinc collocation method. Second, the Hammerstein integral equation arising from chemical reactor theory has been solved by the B-spline wavelet method. Comparison of results has been demonstrated through illustrative examples.

In Chapter 6, the Legendre wavelet method and the Bernoulli wavelet method have been developed to solve a system of integro-differential equations. Legendre wavelets along with their operational matrices are developed to approximate the solutions of a system of nonlinear Volterra integro-differential equations. Also, the nonlinear Volterra weakly singular integro-differential equations system has been solved by the Bernoulli wavelet method. The properties of these wavelets are used to reduce the system of integral equations to a system of algebraic equations which can be solved numerically by Newtons method. Rigorous convergence analysis has been done for these wavelet methods. Illustrative examples have been included to demonstrate the validity and applicability of the proposed techniques.

In Chapter 7, we have solved the second order Lane-Emden type singular differential equation. First, the second order differential equation transforms in to integro-differential equations and is then solved by the Legendre multiwavelet method and the Chebyshev wavelet method. Convergence of these wavelet methods has been discussed in this chapter. In Chapter 8, we have developed an efficient collocation technique called the Legendre spectral collocation method to solve the Fredholm integro-differential-difference equations with variable coefficients and the system of two nonlinear integro-differential equations which arise in biological model. The proposed method is based on the Gauss-Legendre points with the basis functions of Lagrange polynomials. The present method reduces this model to a system of nonlinear algebraic equations and again this algebraic system has been solved numerically by Newton's method.

The study of fuzzy integral equations and fuzzy differential equations is an emerging area of research for many authors. In Chapter 9, we have proposed some numerical techniques for solving fuzzy integral equations and fuzzy integro-differential equations. Fundamentals of fuzzy calculus have been discussed in this chapter. The nonlinear fuzzy Hammerstein integral equation has been solved by Bernstein polynomials and Legendre wavelets, and then compared with the homotopy analysis method. We have solved nonlinear fuzzy Hammerstein Volterra integral equations with constant delay by the Bernoulli wavelet method and then compared with the B-spline wavelet method. Finally, the fuzzy integro-differential equation has been solved by the Legendre wavelet method and compared with the homotopy analysis method. In the fuzzy case, we have applied two-dimensional numerical methods which are discussed in chapter 2. Convergence analysis and error estimate have been provided for the Bernoulli wavelet method.

The study of fractional calculus, fractional differential equations and fractional integral equations has a great importance in the field of science and engineering. Most of the physical phenomena can be modeled by using fractional calculus. Applications of fractional differential equations and fractional integral equations create a wide area of research for many researchers. These works we have placed in Chapter 10. First, the preliminary definitions and theorems of fractional calculus have been presented in this chapter. The nonlinear fractional mixed Volterra-Fredholm integro-differential equations along with mixed boundary conditions have been solved by the Legendre wavelet method. A numerical scheme has been developed by using the Petrov-Galerkin method where the trial and test functions are Legendre wavelets basis functions. Also, this method has been applied to solve fractional Volterra integro-differential equations. The uniqueness and existence of the problem have been discussed and the error estimate of the proposed method has been presented in this work. The Sinc Galerkin method is developed to approximate the solution of fractional Volterra-Fredholm integro-differential equations with weakly singular kernels. The proposed method is based on the Sinc function approximation. The uniqueness and existence of the problem have been discussed and the error analysis of the proposed method has been presented in this chapter.

Brief Biography of Authors

Dr. S. Saha Ray is currently a Professor at the Department of Mathematics, National Institute of Technology, Rourkela, India. Dr. Saha Ray completed his Ph.D. in 2008 from Jadavpur University, Kolkata, India. He received his M.C.A. degree in 2001 from the Indian Institute of Engineering Science and Technology (IIEST), formerly Bengal Engineering College, Shibpur, India. He completed a masters degree in applied mathematics at the Calcutta University, Kolkata, India, in 1998 and a bachelors (honors) degree in mathematics at St. Xaviers College(currently known as St. Xaviers University, Kolkata), Kolkata, India, in 1996. Dr. Saha Ray has about seventeen years of teaching experience at the undergraduate and postgraduate levels in glorious institutes like the National Institute of Technology, Rourkela and two renowned private engineering institutes in Kolkata, West Bengal. He has also about sixteen years of research experience in various fields of applied mathematics. He has published many peer reviewed research papers in numerous fields and various international SCI journals of repute, like *Applied Mathematics and Computation, Communication in Nonlinear Science and Numerical Simulation, Nonlinear Dynamics, Transaction ASME Journal of Applied Mechanics, Journal of Computational and Nonlinear Dynamics, Computers and Mathematics with Applications, Journal of Computational and Applied Mathematics, Mathematical Methods in the Applied Sciences, Computers & Fluids, Physica Scripta, Communications in Theoretical Physics, Nuclear Engineering and Design, International Journal of Nonlinear Science and Numerical Simulation, Annals of Nuclear Energy,* and *Journal of Mathematical Chemistry*, etc. For a detail citation overview, the reader may be referred to Scopus. To date, he has more than 149 research papers published in journals of international repute, including more than 119 SCI journal papers.

He has solely authored a book entitled *Graph Theory with Algorithms and Its Applications: in Applied Science and Technology* published by Springer. A solely authored book entitled *Fractional Calculus with Applications for Nuclear Reactor Dynamics* has been published by the CRC Press of the Taylor & Francis Group. Another solely authored book entitled *Numerical Analysis with Algorithms and Programming* has also been published by the CRC press of Taylor & Francis group, USA. A book entitled *Wavelet Methods for Solving Partial Differential Equations and Fractional Differential Equations* has been recently launched by the CRC press of Taylor & Francis group, USA.

Currently, he is acting as editor-in-chief for the Springer international journal entitled *International Journal of Applied and Computational Mathematics.* He is also an associate editor of a Springer international journal *Mathematical Sciences and Nonlinear Science Letters A* and reviewer of several journals of Elsevier, Springer and Taylor and Francis, etc. He had also been the lead guest editor in the International SCI journals of Hindawi Publishing Corporation, USA.

He has contributed papers on several topics, such as fractional calculus, mathematical modeling, mathematical physics, stochastic modeling, integral equations, and wavelet methods. He is a member of the Society for Industrial and Applied Mathematics (SIAM) and the American Mathematical Society (AMS).

He was the principal investigator of the two *Board of Research in Nuclear Sciences* research projects, with grants from by Bhabha Atomic Research Centre, Mumbai, India. He was also the principal investigator of a research project financed by the *Department of Science and Technology*, Government of India. Currently, he has been acting as principal investigator of another research project financed by the *National Board for Higher Mathematics*, Department of Atomic Energy, Government of India, respectively.

Under his sole supervision, four research scholars had been awarded with Ph.D.s from NIT Rourkela. Currently, he is supervising two research scholars.

It is not out of place to mention that he attended the workshop organized by West Bengal University of Technology (WBUT) on Review of Engineering Degree Curriculum of Mathematics held at National Institute of Technical Teachers Training and Research (NITTTR) Kolkata, from 26th to 30th July 2004. In that workshop the curriculum of mathematics at the undergraduate level as well as the postgraduate level of West Bengal, University of Technology was revised.

He was invited to deliver a lecture in a short term training programme on mathematical modelling organized by the Department of Science, National Institute of Technical Teachers Training and Research (NITTTR) Kolkata, from 4th-8th December 2006. He also attended a summer school on mathematical modelling and its engineering applications organized jointly by the West Bengal State Council of Science & Technology and the Department of Science & Technology, Government of West Bengal in association with B. P. Poddar Institute of Management & Technology, 137, V.I.P. Road, Kolkata-700052, during 27th April to 9th May 2007.

It is worth noting that he was invited to deliver a lecture in a workshop on fractional order systems organized by the Instrumentation & Electronics

Engineering Department, Jadavpur University, Salt Lake Campus, Kolkata sponsored by IEEE Kolkata Chapter, DRDL Hydrabad, BRNS (DAE) Mumbai, from 28th March to 29th March 2008.

Moreover, he was also invited to deliver a lecture in a short term training programme on mathematical modelling organized by the Department of Science, National Institute of Technical Teachers Training and Research (NITTTR) Kolkata, from 24th-28th November 2008.

He was convener of the Symposium on Recent Trends and Emerging Applications of Mathematical Sciences (SRTEAMS-2013) sponsored by CSIR, New Delhi; DST, Govt. of India and INSA, New Delhi held on May 16 to May 17, 2013. He attended as invited participant in the International Conference on Mathematical Modeling in Physical Sciences held in Madrid, Spain from August 28th to August 31st, 2014.

Recently, he attended as invited participant in the Global Conference on Applied Physics and Mathematics held in Rome, Italy from July 25th to July 27th, 2016. The travel grant was sponsored by SERB, DST, Government of India.

Dr. P. K. Sahu is currently working as an Assistant Professor in the Department of Mathematics, School of Applied Sciences, Kalinga Institute of Industrial Technology (KIIT) Deemed to be University, Bhubaneswar, Odisha, India. Dr. Sahu completed his Ph. D. in 2017 from the National Institute of Technology, Rourkela, Odisha. He completed his M.Sc. from Utkal University, Odisha in the year 2011 and graduated from Utkal University with Mathematics Honors in the year 2009. He has research experience of more than 5 years in the field of applied mathematics. He has many papers published in various national and international journals of repute.

Acknowledgement

First, I would like to express my sincere gratitude to my Ph. D. supervisor Prof. Santanu Saha Ray for giving me the guidance and motivation for preparing this book. I gratefully acknowledge the support provided by the National Institute of Technology (NIT), Rourkela and Kalinga Institute of Industrial Technology (KIIT) Deemed to be University, Bhubaneswar. I owe a sense of gratitude to the honorable Founder, KIIT and KISS for his encouraging speeches. I am grateful to all the faculty members and staff of the Mathematics Department for their many helpful comments, encouragement, and sympathetic cooperation. I wish to thank all my colleagues and friends for their encouragement and moral support. Last but not the least, I would like to thank my family: my parents, my wife and my brothers and sister for supporting me spiritually throughout writing this book and my life in general. I thank all those who have ever bestowed upon me their best wishes.

Prakash Kumar Sahu

Chapter 1

Preliminary Concepts

1.1 Introduction

For many years the subject of functional equations has held a prominent place in the attention of mathematicians. In more recent years this attention has been directed to a particular kind of functional equation, an integral equation, wherein the unknown function occurs under the integral sign. Such equations occur widely in diverse areas of applied mathematics and physics. They offer a powerful technique for solving a variety of practical problems. One obvious reason for using the integral equation rather than differential equations is all of the conditions specifying the initial value problem or boundary value problem for a differential equation. In the case of PDEs, the dimension of the problem is reduced in this process so that, for example, a boundary value problem for a practical differential equation in two independent variables transforms into an integral equation involving an unknown function of only one variable. This reduction of what may represent a complicated mathematical model of a physical situation into a single equation is itself a significant step, but there are other advantages to be gained by replacing differentiation with integration. Some of these advantages arise because integration is a smooth process, a feature which has significant implications when approximate solutions are sought. Whether one is looking for an exact solution to a given problem or having to settle for an approximation to it, an integral equation formulation can often provide a useful way forward. For this reason integral equations have attracted attention for most of the last century and their theory is well-developed.

In 1825 Abel, an Italian mathematician, first produced an integral equation in connection with the famous *tautochrone* problem . The problem is connected with the determination of a curve along which a heavy particle, sliding without friction, descends to its lowest position, or more generally, such that the time of descent is a given function of its initial position.

1.2 Integral equation

An integral equation is an equation in which an unknown function appears under one or more integral signs. For example, for $a \leq x \leq b$, $a \leq t \leq b$, the equations

$$\int_a^b K(x,t)y(t)dt = f(x),$$

$$y(x) - \lambda \int_a^b K(x,t)y(t)dt = f(x)$$

$$\text{and} \quad y(x) = \int_a^b K(x,t)[y(t)]^2 dt,$$

where the function $y(x)$, is the unknown function while $f(x)$ and $K(x,t)$ are known functions and λ, a and b are constants, are all integral equations. The above mentioned functions can be real or complex valued functions in x and t. In this work, we have considered only real valued functions.

1.3 Classification of integral equations

An integral equation can be classified as a linear or nonlinear integral equation as similar in the ordinary and partial differential equations. We have noticed that the differential equation can be equivalently represented by the integral equation. Therefore, there is a good relationship between these two equations. An integral equation is called linear if only linear operations are performed in it upon the unknown function. An integral equation which is not linear is known as a nonlinear integral equation. The most frequently used integral equations fall under two major classes, namely Volterra and Fredholm integral equations. Also, we have to classify them as homogeneous or non-homogeneous integral equations.

1.3.1 Fredholm integral equation

The most general form of Fredholm linear integral equations is given by the form

$$g(x)y(x) = f(x) + \lambda \int_a^b K(x,t)y(t)dt, \tag{1.1}$$

where a, b are both constants, $f(x)$, $g(x)$ and $K(x,t)$ are known functions while $y(x)$ is an unknown function and λ is a non-zero real or complex parameter; this is called a Fredholm integral equation of third kind. The $K(x,t)$ is known as the kernel of the integral equation.

- **Fredholm integral equation of the first kind:** A linear integral equation of the form (by setting $g(x) = 0$ in (1.1))

  $$f(x) + \lambda \int_a^b K(x,t)y(t)dt = 0,$$

  is known as a Fredholm integral equation of the first kind.

- **Fredholm integral equation of the second kind:** A linear integral equation of the form (by setting $g(x) = 1$ in (1.1))

  $$y(x) = f(x) + \lambda \int_a^b K(x,t)y(t)dt,$$

  known as a Fredholm integral equation of the second kind.

- **Homogeneous Fredholm integral equation of the second kind:** A linear integral equation of the form (by setting $f(x) = 0, g(x) = 1$ in (1.1))

  $$y(x) = \lambda \int_a^b K(x,t)y(t)dt,$$

  is known as the homogeneous Fredholm integral equation of the second kind.

1.3.2 Volterra integral equation

The most general form of Volterra linear integral equations is given by the form

$$g(x)y(x) = f(x) + \lambda \int_a^x K(x,t)y(t)dt, \tag{1.2}$$

where a is constant, $f(x)$, $g(x)$ and $K(x,t)$ are known functions while $y(x)$ is an unknown function and λ is a non-zero real or complex parameter; this is called a Volterra integral equation of third kind. The $K(x,t)$ is known as the kernel of the integral equation.

- **Volterra integral equation of the first kind:** A linear integral equation of the form (by setting $g(x) = 0$ in (1.2))

$$f(x) + \lambda \int_a^x K(x,t)y(t)dt = 0,$$

is known as a Volterra integral equation of the first kind.

- **Volterra integral equation of the second kind:** A linear integral equation of the form (by setting $g(x) = 1$ in (1.2))

$$y(x) = f(x) + \lambda \int_a^x K(x,t)y(t)dt,$$

is known as a Volterra integral equation of the second kind.

- **Homogeneous Fredholm integral equation of the second kind:** A linear integral equation of the form (by setting $f(x) = 0, g(x) = 1$ in (1.2))

$$y(x) = \lambda \int_a^x K(x,t)y(t)dt,$$

is known as the homogeneous Volterra integral equation of the second kind.

1.3.3 Singular integral equation

A singular integral equation is defined as an integral with infinite limits or when the kernel of the integral becomes unbounded at one or more points within the interval of integration. For example,

$$y(x) = f(x) + \lambda \int_{-\infty}^{\infty} e^{-|x-t|}y(t)dt$$

$$\text{and} \quad f(x) = \int_0^x \frac{1}{(x-t)^\alpha}y(t)dt, 0 < \alpha < 1$$

are singular integral equations.

1.3.4 Integro-differential equation

In the early 1900s, Vito Volterra studied the phenomenon of population growth, and new types of equations have been developed and termed as integro-differential equations. In this type of equation, the unknown function $y(x)$ appears as the combination of the ordinary derivative and under the integral sign. For example,

$$y''(x) = f(x) + \lambda \int_0^x (x-t)y(t)dt, \quad y(0) = 0, y'(0) = 1$$

$$y'(x) = f(x) + \lambda \int_0^1 (xt)y(t)dt, \quad y(0) = 1$$

The above equations are a second order Volterra integro-differential equation and a first order Fredholm integro-differential equation, respectively.

1.3.5 Special kind of kernels

Kernel function is a main part of the integral equation. Classification of kernel functions are as follow:

- **Symmetric kernel:** A kernel $K(x,t)$ is symmetric (or complex symmetric) if
$$K(x,t) = \overline{K}(t,x)$$
where the bar denotes the complex conjugate. A real kernel $K(x,t)$ is symmetric if
$$K(x,t) = K(t,x).$$

- **Separable or degenerate kernel:** A kernel $K(x,t)$ is called separable or degenerate if it can be expressed as the sum of a finite number of terms, each of which is the product of a function of x only and a function of t only, i.e.,
$$K(x,t) = \sum_{i=0}^{n} g_i(x)h_i(t).$$

- **Non-degenerate kernel:** A kernel $K(x,t)$ is called non-degenerate if it cannot be separated as the function of x and function of t. For example, e^{xt}, $\sqrt{x+t}$ are the non-degenerate kernels.

1.4 Wavelets

Wavelets theory is a relatively new and emerging area in mathematical research. It has been applied in a wide range of engineering disciplines; particularly, wavelets are very successfully used in signal analysis for waveform representations and segmentations, time frequency analysis, and fast algorithms for easy implementation [53, 54]. Wavelets permit the accurate representation of a variety of functions and operators. The concept of wavelet analysis has been in place in one form or the other since the beginning of this century. However, in its present form, wavelet theory drew attention in the 1980s with the work of several researchers from various disciplines: Stromberg, Morlet, Grossmann, Meyer, Battle, Lemarie, Coifman, Daubechies, Mallat, and Chui, to name a few. Many other researchers have also made significant contributions.

In applications to discrete data sets, wavelets may be considered basis functions generated by dilations and translations of a single function. Analogous to Fourier analysis, there are wavelet series (WS) and integral wavelet transforms (IWT). In wavelet analysis, WS and IWT are intimately related. Wavelet techniques enable us to divide a complicated function into several simpler ones and study them separately. This property, along with fast wavelet algorithms which are comparable in efficiency to fast Fourier transform algorithms, makes these techniques very attractive for analysis and synthesis. Different types of wavelets have been used as tools to solve problems in signal analysis, image analysis, medical diagnostics, boundary value problems, geophysical signal processing, statistical analysis, pattern recognition, and many others. While wavelets have gained popularity in these areas, new applications are continually being investigated.

Wavelets may be seen as small waves $\psi(t)$, which oscillate at least a few times, but unlike the harmonic waves must die out to zero as $t \to \pm\infty$. The most applicable wavelets are those that die out to identically zero after a few oscillations on a finite interval, i.e., $\psi(t) = 0$ outside the interval. Such a special interval is called the support or compact support of the given (basic) wavelet $\psi(t)$. We say a basic wavelet since it will be equipped with two parameters, namely, scale a and translation b to result in a family of wavelets $\psi\left(\frac{t-b}{a}\right)$. The construction of basic wavelets is established in terms of their associated building blocks or scaling functions $\phi(t)$. The latter is governed by an equation called the recurrence relation or scaling relation. In wavelet analysis, usually, a single scaling function series is used to yield an approximated version of the given signal. Another series of the associated wavelets is added to the former to bring about a refinement. The result is a satisfactory representation of the signal. Once scaling functions are found, it is a simple computation to construct their associated basic wavelets. The scaling functions or building blocks are of paramount importance in the study of wavelet analysis in this chapter.

We consider, in this chapter, the space $L^2(\mathbb{R})$ of measurable functions f, defined on the real line $\mathbb{R}$, that satisfy

$$\int_{-\infty}^{\infty} |f(t)|^2 dt \leq \infty.$$

In fact, we look for such waves that generate $L^2(\mathbb{R})$, these waves should decay to zero at $\pm\infty$; and for all practical purpose, the decay should be very fast.

That is, we look for small waves, or wavelets, to generate $L^2(\mathbb{R})$. For this purpose, we prefer a single function ψ that generates all of $L^2(\mathbb{R})$. Since, ψ is very fast decay, to cover a whole real line, we shift ψ along $\mathbb{R}$. For computational efficiency, we have used integral powers of 2 for frequency partitioning. That is, consider the small waves

$$\psi(2^j t - k), \quad j, k \in \mathbb{Z}.$$

$\psi(2^j t - k)$ is obtained from a single wavelet function $\psi(t)$ by a binary dilation (dilation by 2^j) and a dyadic translation (of $k/2^j$). Any wavelet function $\psi \in L^2(\mathbb{R})$ has two arguments as $\psi_{j,k}$ and defined by

$$\psi_{j,k}(t) = 2^{j/2}\psi(2^j t - k), \quad j, k \in \mathbb{Z},$$

where the quantity $2^{j/2}$ is for normality.

Definition 1.4.1. (Orthogonal wavelet) *A wavelet $\psi \in L^2(\mathbb{R})$ is called an orthogonal wavelet, if the family $\{\psi_{j,k}\}$, is an orthonormal basis of $L^2(\mathbb{R})$; that is,*

$$\langle \psi_{j,k}, \psi_{l,m} \rangle = \delta_{j,l}\delta_{k,m}, \qquad j, k, l, m \in \mathbb{Z}.$$

Definition 1.4.2. (Semi-orthogonal wavelet) *A wavelet $\psi \in L^2(\mathbb{R})$ is called an semi-orthogonal wavelet, if the family $\{\psi_{j,k}\}$ satisfies the following condition,*

$$\langle \psi_{j,k}, \psi_{l,m} \rangle = 0, \quad j \neq l, \qquad j, k, l, m \in \mathbb{Z}.$$

1.4.1 Multiresolution analysis (MRA)

Any wavelet, orthogonal or semi-orthogonal, generates a direct sum decomposition of $L^2(\mathbb{R})$. For each $j \in \mathbb{Z}$, let us consider the closed subspaces

$$V_j = \ldots \oplus W_{j-2} \oplus W_{j-1}, \quad j \in \mathbb{Z}$$

of $L^2(\mathbb{R})$. A set of subspaces $\{V_j\}_{j\in\mathbb{Z}}$ is said to be MRA of $L^2(\mathbb{R})$ if it possesses the following properties:

1. $V_j \subset V_{j+1}, \forall j \in \mathbb{Z}$,
2. $\bigcup_{j\in\mathbb{Z}} V_j$ is dense in $L^2(\mathbb{R})$,
3. $\bigcap_{j\in\mathbb{Z}} V_j = \phi$,
4. $V_{j+1} = V_j \oplus W_j$,
5. $f(t) \in V_j \Leftrightarrow f(2t) \in V_{j+1}, \ \forall j \in \mathbb{Z}$.

Properties (2)-(5) state that $\{V_j\}_{j\in\mathbb{Z}}$ is a nested sequence of subspaces that effectively covers $L^2(\mathbb{R})$. That is, every square integrable function can be approximated as closely as desired by a function that belongs to at least one of the subspaces V_j. A function $\varphi \in L^2(\mathbb{R})$ is called a scaling function if it generates the nested sequence of subspaces V_j and satisfies the dilation equation, namely

$$\phi(t) = \sum_k p_k \phi(at - k), \tag{1.3}$$

with $p_k \in l^2$ and a being any rational number.
For each scale j, since $V_j \subset V_{j+1}$, there exists a unique orthogonal complementary subspace W_j of V_j in V_{j+1}. This subspace $W_j = \overline{Span\{\psi_{j,k} | \psi_{j,k} = \psi(2^j t - k)\}}$ is called wavelet subspace and is generated by $\psi_{j,k} = \psi(2^j t - k)$, where $\psi \in L^2$ is called the wavelet. From the above discussion, these results follow easily:

- $V_{j_1} \bigcap V_{j_2}, j_1 > j_2,$
- $W_{j_1} \bigcap W_{j_2} = 0, j_1 \neq j_2,$
- $V_{j_1} \bigcap W_{j_2} = 0, j_1 \leq j_2.$

Chapter 2

Numerical Methods and Function Approximation

2.1 Introduction

This chapter provides a brief description of the numerical methods for solving linear and nonlinear integral equations, integro-differential equations and systems. Typically, these methods are based on the approximations like wavelets approximations, orthogonal polynomials and orthogonal functions approximations. In this chapter, we introduce the wavelet methods like B-spline wavelet method (BSWM), Legendre wavelet method (LWM), Legendre multi-wavelet method (LMWM), Chebyshev wavelet method (CWM), Haar wavelet method (HWM), Bernoulli wavelet method (BWM), polynomial approximation via Bernstein polynomials, Legendre spectral collocation method, Legendre polynomial, Block-Pulse functions, Sinc functions etc. which are applied to solve different types of integral equations.

2.2 B-spline wavelet method

Wavelets theory is a relatively new and emerging area in mathematical research. It has been applied in a wide range of engineering disciplines; particularly, wavelets are very successfully used in signal analysis for waveform representations and segmentations, time frequency analysis, and fast algorithms for easy implementation [53]. Wavelets permit the accurate representation of a variety of functions and operators. Moreover, wavelets establish a connection with fast numerical algorithms. Wavelets can be separated into two distinct types, orthogonal and semi-orthogonal [53, 68]. The research works available in open literature on integral equation methods have shown a marked preference for orthogonal wavelets [148]. This is probably because the original wavelets, which were widely used for signal processing, were primarily orthogonal. In signal processing applications, unlike integral equation methods, the wavelet itself is never constructed since only its scaling function and coefficients are needed. However, orthogonal wavelets either have infinite support or a non-symmetric, and in some cases fractal, nature. These properties can make them a poor choice for characterization of a function. In contrast, the semi-orthogonal wavelets have finite support, both even and odd symmetry, and simple analytical expressions, ideal attributes of a basis function [148]. We apply compactly supported linear semi-orthogonal B-spline wavelets, specially constructed for the bounded interval to approximate the unknown function present in the integral equations.

2.2.1 B-spline scaling and wavelet functions

Semi-orthogonal wavelets use B-splines specially constructed for the bounded interval and these wavelets can be represented in a closed form. This provides a compact support. Semi-orthogonal wavelets form the basis in the space $L^2(\mathbb{R})$.

Using this basis, an arbitrary function in $L^2(\mathbb{R})$ can be expressed as the wavelet series [53]. For the finite interval $[0, 1]$, the wavelet series cannot be completely presented by using this basis. This is because supports of some basis are truncated at the left or right end points of the interval. Hence a special basis has to be introduced into the wavelet expansion on the finite interval. These functions are referred to as the boundary scaling functions and boundary wavelet functions.

Let m and n be two positive integers and

$$a = x_{-m+1} = ... = x_0 < x_1 < ... < x_n = x_{n+1} = ... = x_{n+m-1} = b, \tag{2.1}$$

be an equally spaced knots sequence. The functions

$$B_{m,j,X}(x) = \frac{x - x_j}{x_{j+m-1} - x_j} B_{m-1,j,X}(x) + \frac{x_{j+m} - x}{x_{j+m} - x_{j+1}} B_{m-1,j+1,X}(x),$$
$$j = -m+1, ..., n-1. \tag{2.2}$$

and

$$B_{1,j,X}(x) = \begin{cases} 1 & , x = [x_j, x_{j+1}) \\ 0 & , otherwise \end{cases} \tag{2.3}$$

are called cardinal B-spline functions of order $m \geq 2$ for the knot sequence $X = \{x_i\}_{i=-m+1}^{n+m-1}$, and

$$Supp\, B_{m,j,X}(x) = [x_j, x_{j+m}] \bigcap [a, b]. \tag{2.4}$$

By considering the interval $[a, b] = [0, 1]$, at any level $j \in \mathbb{Z}^+$, the discretization step is 2^{-j}, and this generates $n = 2^j$ number of segments in $[0, 1]$ with knot sequence

$$X^{(j)} = \begin{cases} x_{-m+1}^{(j)} = ... = x_0^{(j)} = 0, & \\ x_k^{(j)} = \frac{k}{2^j}, & k = 1, ..., n-1, \\ x_n^{(j)} = ... = x_{n+m-1}^{(j)} = 1. & \end{cases} \tag{2.5}$$

Let j_0 be the level for which $2^{j_0} \geq 2m - 1$; for each level, $j \geq j_0$ the scaling functions of order m can be defined as follows in [134]:

$$\varphi_{m,j,i}(x) = \begin{cases} B_{m,j_0,i}(2^{j-j_0}x), & i = -m+1, ..., -1, \\ B_{m,j_0,2^j-m-i}(1 - 2^{j-j_0}x), & i = 2^j - m + 1, ..., 2^j - 1, \\ B_{m,j_0,0}(2^{j-j_0}x - 2^{j_0}i), & i = 0, ..., 2^j - m. \end{cases} \tag{2.6}$$

And the two scale relation for the m-order semi-orthogonal compactly supported B-wavelet functions are defined as follows:

$$\psi_{m,j,i-m} = \sum_{k=i}^{2i+2m-2} q_{i,k} B_{m,j,k-m}, \; i = 1, ..., m-1, \tag{2.7}$$

$$\psi_{m,j,i-m} = \sum_{k=2i-m}^{2i+2m-2} q_{i,k} B_{m,j,k-m}, \; i = m, ..., n-m+1, \tag{2.8}$$

$$\psi_{m,j,i-m} = \sum_{k=2i-m}^{n+i+m-1} q_{i,k} B_{m,j,k-m}, \; i = n-m+2, ..., n, \tag{2.9}$$

where $q_{i,k} = q_{k-2i}$.
Hence there are $2(m-1)$ boundary wavelets and $(n-2m+2)$ inner wavelets in the bounded interval $[a,b]$. Finally, by considering the level j with $j \geq j_0$, the B-wavelet functions in $[0,1]$ can be expressed as follows:

$$\psi_{m,j,i}(x) = \begin{cases} \psi_{m,j_0,i}(2^{j-j_0}x), & i = -m+1, ..., -1, \\ \psi_{m,2^j-2m+1-i,i}(1-2^{j-j_0}x), & i = 2^j-2m+2, ..., 2^j-m, \\ \psi_{m,j_0,0}(2^{j-j_0}x - 2^{-j_0}i), & i = 0, ..., 2^j-2m+1. \end{cases} \tag{2.10}$$

The scaling functions $\varphi_{m,j,i}(x)$ occupy m segments and the wavelet functions $\psi_{m,j,i}(x)$ occupy $2m-1$ segments.

When the semi-orthogonal wavelets are constructed from B-spline of order m, the lowest octave level $j = j_0$ is determined in [69, 108] by

$$2^{j_0} \geq 2m-1, \tag{2.11}$$

so as to have a minimum of one complete wavelet on the interval $[0,1]$.

2.2.2 Function approximation

A function $f(x)$ defined over interval $[0,1]$ may be approximated by B-spline wavelets as [53, 68]

$$f(x) = \sum_{k=-1}^{2^{j_0}-1} c_{j_0,k}\varphi_{j_0,k}(x) + \sum_{j=j_0}^{\infty}\sum_{k=-1}^{2^j-2} d_{j,k}\psi_{j,k}(x). \tag{2.12}$$

In particular, for $j_0 = 2$, if the infinite series in equation (2.12) is truncated at M , then eq. (2.12) can be written as [134, 108]

$$f(x) \approx \sum_{k=-1}^{3} c_k\varphi_{2,k}(x) + \sum_{j=2}^{M}\sum_{k=-1}^{2^j-2} d_{j,k}\psi_{j,k}(x) = C^T\Psi(x). \tag{2.13}$$

where $\varphi_{2,k}$ and $\psi_{j,k}$ are scaling and wavelet functions, respectively, and C and Ψ are $(2^{M+1}+1)\times 1$ vectors given by

$$C = [c_{-1}, c_0, ..., c_3, d_{2,-1}, ..., d_{2,2}, d_{3,-1}, ..., d_{3,6}, ..., d_{M,-1}, ..., d_{M,2^M-2}]^T, \quad (2.14)$$

$$\Psi = [\varphi_{2,-1}, \varphi_{2,0}, ..., \varphi_{2,3}, \psi_{2,-1}, ..., \psi_{2,2}, \psi_{3,-1}, ..., \psi_{3,6}, ..., \psi_{M,-1}, ..., \psi_{M,2^M-2}]^T, \quad (2.15)$$

with

$$c_k = \int_0^1 f(x)\tilde{\varphi}_{2,k}(x)dx, \quad k = -1, 0, ..., 3,$$

$$d_{j,k} = \int_0^1 f(x)\tilde{\psi}_{j,k}(x)dx, \quad j = 2, 3, ..., M, \ k = -1, 0, 1, ..., 2^j - 2, \quad (2.16)$$

where $\tilde{\varphi}_{2,k}(x)$ and $\tilde{\psi}_{j,k}(x)$ are dual functions of $\varphi_{2,k}$ and $\psi_{j,k}$, respectively. These can be obtained by linear combinations of $\varphi_{2,k}, k = -1, ..., 3$ and $\psi_{j,k}, j = 2, ..., M, k = -1, ..., 2^j - 2$, as follows. Let

$$\Phi = [\varphi_{2,-1}(x), \varphi_{2,0}(x), \varphi_{2,1}(x), \varphi_{2,2}(x), \varphi_{2,3}(x)]^T, \quad (2.17)$$

$$\bar{\Psi} = [\psi_{2,-1}(x), \psi_{2,0}(x), ..., \psi_{M,2^M-2}(x)]^T. \quad (2.18)$$

Using eqs. (2.6) and (2.17), we get

$$\int_0^1 \Phi\Phi^T dx = P_1 = \begin{bmatrix} \frac{1}{12} & \frac{1}{24} & 0 & 0 & 0 \\ \frac{1}{24} & \frac{1}{6} & \frac{1}{24} & 0 & 0 \\ 0 & \frac{1}{24} & \frac{1}{6} & \frac{1}{24} & 0 \\ 0 & 0 & \frac{1}{24} & \frac{1}{6} & \frac{1}{24} \\ 0 & 0 & 0 & \frac{1}{24} & \frac{1}{12} \end{bmatrix}, \quad (2.19)$$

and from eqs. (2.10) and (2.18), we have

$$\int_0^1 \bar{\Psi}\bar{\Psi}^T dx = P_2 = \begin{bmatrix} N_{4\times 4} & & & & \\ & \frac{1}{2}N_{8\times 8} & & & \\ & & \cdot & & \\ & & & \cdot & \\ & & & & \frac{1}{2^{M-2}}N_{2^M\times 2^M} \end{bmatrix}, \quad (2.20)$$

where P_1 and P_2 are 5×5 and $(2^{M+1} - 4) \times (2^{M+1} - 4)$ matrices, respectively, and N is a five diagonal matrix given by

$$N = \begin{bmatrix} \frac{2}{27} & \frac{1}{96} & -\frac{1}{864} & 0 & 0 & \cdot & \cdot & \cdot & 0 \\ \frac{1}{96} & \frac{1}{16} & \frac{5}{432} & -\frac{1}{864} & 0 & \cdot & \cdot & \cdot & 0 \\ -\frac{1}{864} & \frac{5}{432} & \frac{1}{16} & \frac{1}{96} & -\frac{1}{864} & \cdot & \cdot & \cdot & 0 \\ \cdot & \cdot & \cdot & \cdot & \cdot & \cdot & & & \cdot \\ \cdot & \cdot & \cdot & \cdot & \cdot & \cdot & \cdot & \cdot & \cdot \\ \cdot & & & \cdot & \cdot & \cdot & \cdot & \cdot & \cdot \\ 0 & \cdot & \cdot & \cdot & -\frac{1}{864} & \frac{5}{432} & \frac{1}{16} & \frac{5}{432} & -\frac{1}{864} \\ 0 & \cdot & \cdot & \cdot & 0 & -\frac{1}{864} & \frac{5}{432} & \frac{1}{16} & \frac{1}{96} \\ 0 & \cdot & \cdot & \cdot & 0 & 0 & -\frac{1}{864} & \frac{1}{96} & \frac{2}{27} \end{bmatrix}. \tag{2.21}$$

Suppose $\tilde{\Phi}$ and $\tilde{\bar{\Psi}}$ are the dual functions of Φ and $\bar{\Psi}$, respectively, given by

$$\tilde{\Phi} = [\tilde{\phi}_{2,-1}(x), \tilde{\phi}_{2,0}(x), \tilde{\phi}_{2,1}(x), \tilde{\phi}_{2,2}(x), \tilde{\phi}_{2,3}(x)]^T, \tag{2.22}$$

$$\tilde{\bar{\Psi}} = [\tilde{\psi}_{2,-1}(x), \tilde{\psi}_{2,0}(x), ..., \tilde{\psi}_{M,2^M-2}(x)]^T. \tag{2.23}$$

And combining the above two, we can get

$$\tilde{\Psi} = [\tilde{\phi}_{2,-1}(x), \tilde{\phi}_{2,0}(x), ..., \tilde{\phi}_{2,3}(x), \tilde{\psi}_{2,-1}(x), \tilde{\psi}_{2,0}(x), ..., \tilde{\psi}_{M,2^M-2}(x)]^T. \tag{2.24}$$

Using eqs. (2.17)-(2.18) and (2.22)-(2.23), we have

$$\int_0^1 \tilde{\Phi}\Phi^T dx = I_1, \quad \int_0^1 \tilde{\bar{\Psi}}\bar{\Psi}^T dx = I_2, \tag{2.25}$$

where I_1 and I_2 are 5×5 and $(2^{M+1} - 4) \times (2^{M+1} - 4)$ identity matrices, respectively. Then eqs. (2.19), (2.20) and (2.25) yield

$$\tilde{\Phi} = P_1^{-1}\Phi, \quad \tilde{\bar{\Psi}} = P_2^{-1}\bar{\Psi}. \tag{2.26}$$

2.3 Legendre wavelet method

2.3.1 Properties of Legendre wavelets

Wavelets constitute a family of functions constructed from dilation and translation of a single function called a mother wavelet. When the dilation parameter a and the translation parameter b vary continuously, we have the following family of continuous wavelets as

$$\Psi_{a,b}(t) = |a|^{-\frac{1}{2}} \Psi\left(\frac{t-b}{a}\right), \quad a, b \in \mathbb{R}, \quad a \neq 0 \tag{2.27}$$

If we restrict the parameters a and b to discrete values as $a = a_0^{-k}$, $b = nb_0a_0^{-k}$, $a_0 > 1$, $b_0 > 0$ and n, and k are positive integers, from eq. (2.27) we have the following family of discrete wavelets:

$$\psi_{k,n}(t) = |a_0|^{\frac{k}{2}}\psi(a_0^k t - nb_0),$$

where $\psi_{k,n}(t)$ form a wavelet basis for $L^2(\mathbb{R})$. In particular, when $a_0 = 2$ and $b_0 = 1$, then $\psi_{k,n}(t)$ form an orthonormal basis.

Legendre wavelets $\psi_{n,m}(t) = \psi(k, \hat{n}, m, t)$ have four arguments. $\hat{n} = 2n - 1$, $n = 1, 2, ..., 2^{k-1}$, $k \in \mathbb{Z}^+$, m is the order of Legendre polynomials and t is normalized time. They are defined on $[0, 1)$ as

$$\psi_{n,m}(t) = \psi(k, \hat{n}, m, t) = \begin{cases} \sqrt{m + \frac{1}{2}}2^{\frac{k}{2}}P_m(2^k t - \hat{n}), & \frac{\hat{n}-1}{2^k} \le t < \frac{\hat{n}+1}{2^k}, \\ 0, & otherwise, \end{cases} \tag{2.28}$$

where $m = 0, 1, ..., M - 1$ and $n = 1, 2, ..., 2^{k-1}$. The coefficient $\sqrt{m + \frac{1}{2}}$ is for orthonormality, the dilation parameter is $a = 2^{-k}$ and translation parameter is $b = \hat{n}2^{-k}$.

Here $P_m(t)$ are Legendre polynomials of order m, which are orthogonal with respect to weight function $w(t) = 1$ on the interval $[-1, 1]$. This can be determined from the following recurrence formulae:

$$\begin{aligned} &P_0(t) = 1, \\ &P_1(t) = t, \\ &P_{m+1}(t) = \left(\frac{2m+1}{m+1}\right)tP_m(t) - \left(\frac{m}{m+1}\right)P_{m-1}(t), \quad m = 1, 2, 3, ... \end{aligned}$$

The two-dimensional Legendre wavelets are defined as [29, 155]

$$\psi_{n_1,m_1,n_2,m_2}(x,t) = \begin{cases} AP_{m_1}(2^{k_1}x - 2n_1 + 1)P_{m_2}(2^{k_2}t - 2n_2 + 1), \\ \qquad \frac{n_1-1}{2^{k_1-1}} \le x < \frac{n_1}{2^{k_1-1}}, \frac{n_2-1}{2^{k_2-1}} \le t < \frac{n_2}{2^{k_2-1}}, \\ 0, \quad \text{Otherwise}, \end{cases} \tag{2.29}$$

where $A = \sqrt{(m_1 + \frac{1}{2})(m_2 + \frac{1}{2})}2^{\frac{k_1+k_2}{2}}$, and $n_1 = 1, 2, ..., 2^{k_1-1}$, $n_2 = 1, 2, ..., 2^{k_2-1}$, $k_1, k_2 \in \mathbb{Z}^+$, and m_1, m_2 are the order of Legendre polynomials. $\psi_{n_1,m_1,n_2,m_2}(x,t)$ forms a basis for $L^2([0, 1) \times [0, 1))$.

2.3.2 Function approximation by Legendre wavelets

A function $f(x)$ defined over $[0,1)$ can be expressed by the Legendre wavelets as

$$f(x) = \sum_{n=1}^{\infty}\sum_{m=0}^{\infty} c_{n,m}\psi_{n,m}(x), \tag{2.30}$$

where $c_{n,m} = \langle f(x), \psi_{n,m}(x)\rangle$, in which $\langle .,.\rangle$ denotes the inner product. If the infinite series in eq. (2.30) is truncated, then eq. (2.30) can be written as

$$f(x) \cong \sum_{n=1}^{2^{k-1}}\sum_{m=0}^{M-1} c_{n,m}\psi_{n,m}(x) = C^T\Psi(x), \tag{2.31}$$

where C and $\Psi(x)$ are $\left(2^{k-1}M \times 1\right)$ matrices given by

$$C = \left[c_{1,0}, c_{1,1}, ..., c_{1,M-1}, c_{2,0}, ..., c_{2,M-1}, ..., c_{2^{k-1},0}, ..., c_{2^{k-1},M-1}\right]^T, \tag{2.32}$$

$$\Psi(x) = \left[\psi_{1,0}(x), \psi_{1,1}(x), ..., \psi_{1,M-1}(x), ..., \psi_{2^{k-1},0}(x), ..., \psi_{2^{k-1},M-1}(x)\right]^T. \tag{2.33}$$

A function $f(x,t)$ defined over $[0,1) \times [0,1)$ can be expanded by two dimensional Legendre wavelets as [29]

$$f(x,t) = \sum_{n_1=1}^{\infty}\sum_{m_1=0}^{\infty}\sum_{n_2=1}^{\infty}\sum_{m_2=0}^{\infty} c_{n_1,m_1,n_2,m_2}\psi_{n_1,m_1,n_2,m_2}(x,t). \tag{2.34}$$

If the infinite series in Eq. (2.34) is truncated, then it can be written as

$$\begin{aligned} f(x,t) &= \sum_{n_1=1}^{2^{k_1-1}}\sum_{m_1=0}^{M_1-1}\sum_{n_2=1}^{2^{k_2-1}}\sum_{m_2=0}^{M_2-1} c_{n_1,m_1,n_2,m_2}\psi_{n_1,m_1,n_2,m_2}(x,t) \\ &= C^T\Psi(x,t), \end{aligned} \tag{2.35}$$

where C and $\Psi(x,t)$ are $\left(2^{k_1-1}2^{k_2-1}M_1M_2 \times 1\right)$ matrices given by

$$\begin{aligned} C =& [c_{1,0,1,0}, ..., c_{1,0,1,M_2-1}, \cdots \\ & \cdots, , c_{1,0,2^{k_2-1},M_2-1}, \cdots \\ & \cdots, c_{2^{k_1-1},M_1-1,2^{k_2-1},M_2-1}]^T, \end{aligned} \tag{2.36}$$

$$\begin{aligned} \Psi(x,t) =& [\psi_{1,0,1,0}(x,t), ..., \psi_{1,0,1,M_2-1}(x,t), ... \\ & ..., \psi_{1,0,2^{k_2-1},M_2-1}(x,t), ... \\ & ..., \psi_{2^{k_1-1},M_1-1,2^{k_2-1},M_2-1}(x,t)]^T. \end{aligned} \tag{2.37}$$

2.3.3 Operational matrix of derivative by Legendre wavelets

The operational matrix of derivative for Legendre wavelets, D can be defined as follow

$$
\begin{aligned}
\Psi'(x) &= D\Psi(x), \\
\Psi''(x) &= D^2\Psi(x), \\
&. \\
&. \\
&. \\
\Psi^{(p)}(x) &= D^p\Psi(x),
\end{aligned}
$$

where

$$
D = \begin{bmatrix} H & 0 & 0 & . & 0 \\ 0 & H & 0 & . & 0 \\ 0 & 0 & H & . & 0 \\ . & . & . & . & . \\ 0 & 0 & 0 & . & H \end{bmatrix} \tag{2.38}
$$

in which H is $(M+1)\times(M+1)$ matrix and each element is defined as follows

$$
H_{r,s} = \begin{cases} 2^{k+1}\sqrt{(2r-1)(2s-1)}, & r = 2, ..., (M+1),\ s = 1, ..., r-1, \\ & \text{and } (r+s) \text{ odd}, \\ 0, & \text{otherwise.} \end{cases}
$$

2.3.4 Operational matrix of integration by Legendre wavelets

The operational matrix of integration for Legendre wavelets Q can be defined as follow

$$
\begin{aligned}
&\int_0^x \Psi(x)dx = Q\Psi(x), \\
&\int_0^x \int_0^x \Psi(x)dxdx = Q^2\Psi(x), \\
&. \\
&. \\
&. \\
&\int_0^x \int_0^x \cdots \int_0^x \Psi(x)\underbrace{dxdx...dx}_{p-times} = Q^p\Psi(x),
\end{aligned}
$$

where Q is a $\left(2^{k-1}M \times 2^{k-1}M\right)$ matrix and defined as follow

$$Q = \frac{1}{2^k}\begin{bmatrix} X & W & W & . & W \\ 0 & X & W & . & W \\ 0 & 0 & X & . & W \\ . & . & . & . & . \\ 0 & 0 & 0 & . & X \end{bmatrix} \tag{2.39}$$

where X and W are $(M \times M)$ matrices defined as

$$W = \begin{bmatrix} 2 & 0 & 0 & . & 0 \\ 0 & 0 & 0 & . & 0 \\ 0 & 0 & 0 & . & 0 \\ . & . & . & . & . \\ 0 & 0 & 0 & . & 0 \end{bmatrix}$$

and

$$X = \begin{bmatrix} 1 & \frac{1}{\sqrt{3}} & 0 & 0 & . & . & 0 & 0 \\ -\frac{1}{\sqrt{3}} & 0 & \frac{1}{\sqrt{3}\sqrt{5}} & 0 & . & . & 0 & 0 \\ 0 & -\frac{1}{\sqrt{3}\sqrt{5}} & 0 & \frac{1}{\sqrt{5}\sqrt{7}} & . & . & 0 & 0 \\ . & . & . & . & . & & . & . \\ . & . & . & . & . & & . & . \\ . & . & . & . & & & . & . \\ 0 & 0 & 0 & 0 & . & . & 0 & \frac{1}{\sqrt{2M-3}\sqrt{2M-1}} \\ 0 & 0 & 0 & 0 & . & . & -\frac{1}{\sqrt{2M-1}\sqrt{2M-3}} & 0 \end{bmatrix}.$$

2.3.5 Properties of Legendre multi-wavelets

Legendre multi-wavelets are very similar to Legendre wavelets. The Legendre multi-wavelets are defined on the interval $[0, T]$ where T is any positive integer. Legendre multi-wavelets $\psi_{n,m}(t) = \psi(k, n, m, t)$ have four arguments. $n = 0, 1, 2, ..., 2^k - 1$, $k \in \mathbb{Z}^+$, m is the order of Legendre polynomials and t is normalized time. They are defined on $[0, T)$ as [218]

$$\psi_{n,m}(t) = \begin{cases} \sqrt{2m+1}\left(\frac{2^{\frac{k}{2}}}{\sqrt{T}}\right) P_m\left(\frac{2^k t}{T} - n\right), & \frac{nT}{2^k} \leq t < \frac{(n+1)T}{2^k}, \\ 0, & otherwise, \end{cases} \tag{2.40}$$

where $m = 0, 1, ..., M-1$ and $n = 0, 1, 2, ..., 2^k - 1$.

The coefficient $\sqrt{2m+1}$ is for the orthonormality , the dilation parameter is $a = 2^{-k}T$ and the translation parameter is $b = n2^{-k}T$.

Here $P_m(t)$ are the well-known shifted Legendre polynomials of order m, which are defined on the interval $[0, 1]$, and can be determined with the aid of the following recurrence formulae:

$$P_0(t) = 1,$$
$$P_1(t) = 2t - 1,$$
$$P_{m+1}(t) = \left(\frac{2m+1}{m+1}\right)(2t-1)P_m(t) - \left(\frac{m}{m+1}\right)P_{m-1}(t), \quad m = 1, 2, 3, ...$$

A function $f(t)$ defined over $[0, T)$ can be expressed by the Legendre multi-wavelets as

$$f(t) = \sum_{n=0}^{\infty}\sum_{m=0}^{\infty} c_{n,m}\psi_{n,m}(t), \tag{2.41}$$

where $c_{n,m} = \langle f(t), \psi_{n,m}(t)\rangle$, in which $\langle .,. \rangle$ denotes the inner product. If the infinite series in eq. (2.41) is truncated, then eq. (2.41) can be written as

$$f(t) \cong \sum_{n=0}^{2^k-1}\sum_{m=0}^{M} c_{n,m}\psi_{n,m}(t) = C^T\Psi(t), \tag{2.42}$$

where C and $\Psi(t)$ are $\left(2^k(M+1) \times 1\right)$ matrices given by

$$C = \left[c_{0,0}, c_{0,1}, ..., c_{0,M}, c_{1,0}, ..., c_{1,M}, ..., c_{2^k-1,0}, ..., c_{2^k-1,M}\right]^T, \tag{2.43}$$

$$\Psi(t) = \left[\psi_{0,0}(t), \psi_{0,1}(t), ..., \psi_{0,M}(t), ..., \psi_{2^k-1,0}(t), ..., \psi_{2^k-1,M}(t)\right]^T. \tag{2.44}$$

2.4 Chebyshev wavelets method

2.4.1 Properties of Chebyshev wavelets

Wavelets constitute a family of functions constructed from dilation and translation of a single function called a mother wavelet. When the dilation parameter a and the translation parameter b vary continuously, we have the following family of continuous wavelets as

$$\Psi_{a,b}(t) = |a|^{-\frac{1}{2}}\Psi\left(\frac{t-b}{a}\right), \qquad a, b \in \mathbb{R}, \quad a \neq 0. \tag{2.45}$$

If we restrict the parameters a and b to discrete values as $a = a_0^{-k}$, $b =$

$nb_0a_0^{-k}$, $a_0 > 1$, $b_0 > 0$ and n, and k are positive integers, we have the following family of discrete wavelets:

$$\psi_{k,n}(t) = |a_0|^{\frac{k}{2}}\psi(a_0^k t - nb_0), \quad n, k \in \mathbb{Z}^+,$$

where $\psi_{k,n}(t)$ forms a wavelet basis for $L^2(R)$. In particular, when $a_0 = 2$ and $b_0 = 1$, then $\psi_{k,n}(t)$ form an orthonormal basis.

Chebyshev wavelets $\psi_{n,m}(t) = \psi(k, n, m, t)$ have four arguments, where $n = 1, 2, ..., 2^{k-1}$, $k \in \mathbb{Z}^+$, m is the order of Chebyshev polynomials and t is normalized time. They are defined on the interval $[0, 1)$ as [230]

$$\psi_{n,m}(t) = \begin{cases} 2^{\frac{k}{2}}\sqrt{\frac{2}{\pi}}U_m\left(2^k t - 2n + 1\right), & \frac{n-1}{2^{k-1}} \leq t < \frac{n}{2^{k-1}}, \\ 0, & otherwise, \end{cases} \tag{2.46}$$

where $m = 0, 1, ..., M - 1$ and $n = 1, 2, ..., 2^{k-1}$.

The coefficient $\sqrt{\frac{2}{\pi}}$ is for the orthonormality, the dilation parameter is $a = 2^{-k}$ and the translation parameter is $b = (2n - 1)2^{-k}$.

Here $U_m(t)$ are the well-known m^{th} order second Chebyshev polynomials with respect to weight function $w(t) = \sqrt{1 - t^2}$ which are defined on the interval $[-1, 1]$, and can be determined with the aid of the following recurrence formulae

$$\begin{aligned} &U_0(t) = 1, \\ &U_1(t) = 2t, \\ &U_{m+1}(t) = 2tU_m(t) - U_{m-1}(t), \quad m = 1, 2, 3, ... \end{aligned}$$

We should note that in dealing with the second Chebyshev wavelets the weight function $\tilde{w}(t) = w(2t-1)$ has to dilate and translate as $w_n(t) = w(2^k t - 2n + 1)$.

Chebyshev polynomials of second kind are orthogonal with respect to the weight $\sqrt{1 - t^2}$ on the interval $[-1, 1]$ and defined as

$$\int_{-1}^{1} U_m(t)U_n(t)\sqrt{1 - t^2}dt = \begin{cases} 0, & m \neq n, \\ \pi/2, & m = n. \end{cases}$$

2.4.2 Function approximation by Chebyshev wavelets

A function $u(x)$ defined over $[0, 1)$ can be expressed by the Chebyshev wavelets as

$$u(x) = \sum_{n=1}^{\infty}\sum_{m=0}^{\infty} c_{n,m}\psi_{n,m}(x), \tag{2.47}$$

where

$$c_{n,m} = \langle u(x), \psi_{n,m}(x)\rangle_{w_n(x)} = \int_0^1 w_n(x)u(x)\psi_{n,m}(x)dx, \tag{2.48}$$

in which $\langle .,. \rangle$ denotes the inner product in $L^2_{w_n(x)}[0,1]$. If the infinite series in eq. (2.47) is truncated, then the eq. (2.47) can be written as

$$u(x) \cong \sum_{n=1}^{2^{k-1}} \sum_{m=0}^{M-1} c_{n,m}\psi_{n,m}(x) = C^T\Psi(x), \tag{2.49}$$

where, C and $\Psi(x)$ are $(2^{k-1}M \times 1)$ matrices given by

$$C = \left[c_{1,0}, c_{1,1}, ..., c_{1,M-1}, c_{2,0}, ..., c_{2,M-1}, ..., c_{2^{k-1},0}, ..., c_{2^{k-1},M-1}\right]^T, \tag{2.50}$$

$$\Psi(x) = \left[\psi_{1,0}(x), \psi_{1,1}(x), ..., \psi_{1,M-1}(x), ..., \psi_{2^{k-1},0}(x), ..., \psi_{2^{k-1},M-1}(x)\right]^T. \tag{2.51}$$

Lemma 2.4.1. (Bessel's inequality) *Let H be a Hilbert space and suppose that $e_1, e_2, e_3, ...$ is an orthonormal sequence on H; then for any x in H*

$$\sum_{l=1}^{\infty} |\langle x, e_l\rangle|^2 \leq \|x\|^2,$$

where $\langle .,. \rangle$ denotes the inner product in the Hilbert space H.

Theorem 2.4.2. *Let $u(x) \in L^2_w(\mathbb{R})$, the series solution*

$$u(x) = \sum_{n=1}^{\infty}\sum_{m=0}^{\infty} c_{n,m}\psi_{n,m}(x)$$

defined in eq. (2.47) using Chebyshev wavelet method converges to $u(x)$.

Proof. Let $u(x) \in L^2_w(\mathbb{R})$ and $L^2_w(\mathbb{R})$ be the Hilbert space and $\psi_{n,m}$ defined in (2.46) form an orthonormal basis with respect to the weight function $w(x) = \sqrt{1-x^2}$.
Let $u(x) = \sum_{m=0}^{M-1} c_{n,m}\psi_{n,m}(x)$ where $c_{n,m} = \langle u(x), \psi_{n,m}(x)\rangle_{w(x)}$.
Let us denote $\psi_{n,m}(x) = \psi_j(x)$ for a fixed n and let $\alpha_j = \langle u(x), \psi_j(x)\rangle$.
Now, we define the sequence of partial sum $\{S_n\}$ of $\{\alpha_j\psi_j(x)\}_{j\geq 1}$; let $S_n = \sum_{j=1}^{n} \alpha_j\psi_j(x)$.

Now

$$\langle u(x), S_n\rangle = \langle u(x), \sum_{j=1}^{n} \alpha_j\psi_j(x)\rangle = \sum_{j=1}^{n} |\alpha_j|^2.$$

We claim that

$$\|S_n - S_m\|^2 = \sum_{j=m+1}^{n} |\alpha_j|^2, \quad n > m.$$

Now

$$\|\sum_{j=m+1}^{n} \alpha_j\psi_j(x)\|^2 = \langle \sum_{j=m+1}^{n} \alpha_j\psi_j(x), \sum_{j=m+1}^{n} \alpha_j\psi_j(x)\rangle = \sum_{j=m+1}^{n} |\alpha_j|^2, \quad \text{for} \quad n > m.$$

Therefore,

$$\|\sum_{j=m+1}^{n} \alpha_j\psi_j(x)\|^2 = \sum_{j=m+1}^{n} |\alpha_j|^2, \quad for \quad n > m.$$

From Bessel's inequality, we have $\sum_{j=1}^{\infty} |\alpha_j|^2$ is convergent and hence

$$\|\sum_{j=m+1}^{n} \alpha_j\psi_j(x)\|^2 \to 0 \quad \text{as} \quad n \to \infty.$$

So,

$$\|\sum_{j=m+1}^{n} \alpha_j\psi_j(x)\| \to 0$$

and $\{S_n\}$ is a Cauchy sequence and it converges to s (say).
We assert that $u(x) = s$.
Now

$$\begin{aligned}\langle s - u(x), \psi_j(x)\rangle =& \langle s, \psi_j(x)\rangle - \langle u(x), \psi_j(x)\rangle \\ =& \langle \lim_{n\to\infty} S_n, \psi_j(x)\rangle - \alpha_j \\ =& \alpha_j - \alpha_j\end{aligned}$$

This implies

$$\langle s - u(x), \psi_j(x)\rangle = 0$$

Hence $u(x) = s$ and $\sum_{j=1}^{n} \alpha_j\psi_j(x)$ converges to $u(x)$ as $n \to \infty$. □

Lemma 2.4.3. *If the Chebyshev wavelet expansion of a continuous function $u(x)$ converges uniformly, then the Chebyshev wavelet expansion converges to the function $u(x)$.*

Proof. (see reference [5], Lemma 3.1) □

Theorem 2.4.4. *A function $u(x) \in L^2_w([0,1])$, with bounded second derivative, say $|u''(x)| \leq N$, can be expanded as an infinite sum of Chebyshev wavelets, and the series converges uniformly to $u(x)$, that is,*

$$u(x) = \sum_{n=1}^{\infty}\sum_{m=0}^{\infty} c_{n,m}\psi_{n,m}(x).$$

Proof. From eq. (2.48), it follows that

$$\begin{aligned} c_{n,m} &= \int_0^1 u(x)\psi_{n,m}(x)w_n(x)dx \\ &= 2^{k/2}\sqrt{\frac{2}{\pi}}\int_{(n-1)/2^{k-1}}^{n/2^{k-1}} u(x)U_m(2^k x - 2n + 1)w(2^k x - 2n + 1)dx \end{aligned}$$

If $m > 1$, by substituting $2^k x - 2n + 1 = \cos\theta$, we have

$$\begin{aligned} c_{n,m} &= \frac{1}{2^{(k-1)/2}\sqrt{\pi}}\int_0^\pi u\left(\frac{\cos\theta + 2n - 1}{2^k}\right)\sin(m+1)\theta\sin\theta d\theta \\ &= \frac{1}{2^{(k+1)/2}\sqrt{\pi}}\int_0^\pi u\left(\frac{\cos\theta + 2n - 1}{2^k}\right)(\cos m\theta - \cos(m+2)\theta)\, d\theta \\ &= \frac{1}{2^{(k+1)/2}\sqrt{\pi}} u\left(\frac{\cos\theta + 2n - 1}{2^k}\right)\left[\frac{\sin m\theta}{m} - \frac{\sin(m+2)\theta}{m+2}\right]_0^\pi \\ &+ \frac{1}{2^{(3k+1)/2}\sqrt{\pi}}\int_0^\pi u'\left(\frac{\cos\theta + 2n - 1}{2^k}\right) \\ &\quad \sin\theta\left[\frac{\sin m\theta}{m} - \frac{\sin(m+2)\theta}{m+2}\right] d\theta \\ &\quad (\textit{integrating by parts}) \\ &= \frac{1}{2^{(5k+1)/2}\sqrt{\pi}}\int_0^\pi u''\left(\frac{\cos\theta + 2n - 1}{2^k}\right)\gamma_m(\theta)d\theta \\ &\quad (\textit{again integrating by parts}) \end{aligned}$$

where

$$\begin{aligned} \gamma_m(\theta) = &\frac{\sin\theta}{m}\left(\frac{\sin(m-1)\theta}{m-1} - \frac{\sin(m+1)\theta}{m+1}\right) \\ &- \frac{\sin\theta}{m+2}\left(\frac{\sin(m+1)\theta}{m+1} - \frac{\sin(m+3)\theta}{m+3}\right) \end{aligned}$$

Therefore,

$$\begin{aligned}
|c_{n,m}| &= |\frac{1}{2^{(5k+1)/2}\sqrt{\pi}}\int_0^{\pi} u''\left(\frac{\cos\theta+2n-1}{2^k}\right)\gamma_m(\theta)d\theta| \\
&= \frac{1}{2^{(5k+1)/2}\sqrt{\pi}}|\int_0^{\pi} u''\left(\frac{\cos\theta+2n-1}{2^k}\right)\gamma_m(\theta)d\theta| \\
&\leq \frac{1}{2^{(5k+1)/2}\sqrt{\pi}}\int_0^{\pi} |\gamma_m(\theta)|d\theta \\
&\leq \frac{N\sqrt{\pi}}{2^{(5k+1)/2}}\left[\frac{1}{m}\left(\frac{1}{m-1}+\frac{1}{m+1}\right)+\frac{1}{m+2}\left(\frac{1}{m+1}+\frac{1}{m+3}\right)\right] \\
&= \frac{N\sqrt{\pi}}{2^{(5k+1)/2}}\left(\frac{4}{m^2+2m-3}\right) \\
&= \frac{N\sqrt{\pi}}{2^{(5k-3)/2}(m^2+2m-3)}
\end{aligned}$$

Since $n < 2^{k-1}$, we obtain

$$|c_{n,m}| < \frac{N\sqrt{\pi}}{2n^{5/2}(m^2+2m-3)}.$$

Now, if $m = 1$, we have

$$|c_{n,1}| < \frac{2\sqrt{\pi}}{2^{(3k+1)/2}}\max_{0\leq x\leq 1}|u'(x)|.$$

Hence the series $\sum_{n=1}^{\infty}\sum_{m=1}^{\infty} c_{n,m}$ is absolutely convergent.

For $m = 0$, $\{\psi_{n,0}\}_{n=1}^{\infty}$ form an orthogonal system constructed by the Haar scaling function with respect to the weight function $w(x)$, and thus $\sum_{n=0}^{\infty} c_{n,0}\psi_{n,0}(x)$ is convergent.

On the other hand, we have

$$\begin{aligned}
|\sum_{n=1}^{\infty}\sum_{m=0}^{\infty} c_{n,m}\psi_{n,m}(x)| &\leq |\sum_{n=1}^{\infty} c_{n,0}\psi_{n,0}(x)| + \sum_{n=1}^{\infty}\sum_{m=1}^{\infty}|c_{n,m}||\psi_{n,m}(x)| \\
&\leq |\sum_{n=1}^{\infty} c_{n,0}\psi_{n,0}(x)| + \sum_{n=1}^{\infty}\sum_{m=1}^{\infty}|c_{n,m}| < \infty.
\end{aligned}$$

Therefore, utilizing Lemma 2.4.3, the series $\sum_{n=1}^{\infty}\sum_{m=0}^{\infty} c_{n,m}\psi_{n,m}(x)$ converges to $u(x)$ uniformly. □

2.5 Haar wavelet method

2.5.1 Properties of Haar wavelets

Haar functions have been used from 1910 when they were introduced by the Hungarian mathematician Alferd Haar. Haar wavelets are the simplest wavelets among various types of wavelets. They are step functions (piecewise constant functions) on the real line that can take only three values, i.e., 0, 1 and -1. We use the Haar wavelet method due to the following features: it is simpler and fast, flexible, convenient, has small computational costs and is computationally attractive.

The Haar functions are the family of switched rectangular waveforms where amplitudes can differ from one function to another function. Usually the Haar wavelets are defined on the interval $[0, 1]$ but in general cases these are defined on $[a, b]$. We divide the interval $[a, b]$ into m equal subintervals. In this case the orthogonal set of Haar functions is defined on the interval $[a, b]$ by [172, 175]

$$h_0(x) = \begin{cases} 1, & x \in [a, b] \\ 0, & \text{otherwise} \end{cases} \tag{2.52}$$

and

$$h_i(x) = \begin{cases} 1, & \xi_1(i) \leq x < \xi_2(i) \\ -1, & \xi_2(i) \leq x < \xi_3(i) \\ 0, & \text{otherwise} \end{cases} \tag{2.53}$$

where

$$\xi_1(i) = a + \frac{k-1}{2^j}(b-a),$$

$$\xi_2(i) = a + \frac{k - \frac{1}{2}}{2^j}(b-a),$$

$$\xi_3(i) = a + \frac{k}{2^j}(b-a),$$

where $i = 1, 2, ..., m-1$, $m = 2^J$, J is the positive integer, called the maximum level of resolution. Here j and k represent the integer decomposition of the index i. i.e. $i = k + 2^j - 1$, $0 \leq j < i$ and $1 \leq k < 2^j + 1$.

Mutual orthogonalities of all Haar wavelets can be expressed as

$$\int_a^b h_m(x)h_n(x)dx = (b-a)2^{-j}\delta_{mn} = \begin{cases} (b-a)2^{-j}, & m = n = 2^j + k, \\ 0, & m \neq n. \end{cases} \tag{2.54}$$

2.5.2 Function approximation by Haar wavelets

An arbitrary function $y(x) \in L^2[a, b]$ can be expanded into the following Haar series

$$y(x) = \sum_{i=0}^{+\infty} c_i h_i(x), \tag{2.55}$$

where the coefficients c_i are given by

$$c_i = (b-a)^{-1} 2^j \int_a^b y(x) h_i(x) dx. \tag{2.56}$$

Eq. (2.55) can be approximated with finite terms as follows:

$$y(x) \approx \sum_{i=0}^{m-1} c_i h_i(x) = C_{(m)}^T h_{(m)}(x), \tag{2.57}$$

where the coefficient vector $C_{(m)}^T$ and the Haar function vector $h_{(m)}(x)$ are, respectively, defined as

$$\begin{aligned} C_{(m)}^T &= [c_0, c_1, ..., c_{m-1}], \\ h_{(m)}(x) &= [h_0(x), h_1(x), ..., h_{m-1}(x)]^T. \end{aligned} \tag{2.58}$$

2.6 Bernoulli wavelet method

2.6.1 Properties of Bernoulli wavelets

Wavelets constitute a family of functions constructed from dilation and translation of a single function called a mother wavelet. When the dilation parameter a and the translation parameter b vary continuously, we have the following family of continuous wavelets as

$$\Psi_{a,b}(t) = |a|^{-\frac{1}{2}} \Psi\left(\frac{t-b}{a}\right), \qquad a, b \in \mathbb{R}, \quad a \neq 0. \tag{2.59}$$

If we restrict the parameters a and b to discrete values as $a = a_0^{-k}$, $b = nb_0 a_0^{-k}$, $a_0 > 1$, $b_0 > 0$ and n, and k are positive integers, we have the following family of discrete wavelets:

$$\psi_{k,n}(t) = |a_0|^{\frac{k}{2}} \psi(a_0^k t - nb_0), \quad n, k \in \mathbb{Z}^+,$$

where $\psi_{k,n}(t)$ forms a wavelet basis for $L^2(\mathbb{R})$. In particular, when $a_0 = 2$ and $b_0 = 1$, then $\psi_{k,n}(t)$ form an orthonormal basis.

Bernoulli wavelets $\psi_{n,m}(t) = \psi(k, n, m, t)$ have four arguments, where $n = 1, 2, ..., 2^{k-1}$, $k \in \mathbb{Z}^+$, m is the order of Bernoulli polynomials and t is normalized time. They are defined on the interval $[0, 1)$ as [96]

$$\psi_{n,m}(t) = \begin{cases} 2^{\frac{k-1}{2}} \tilde{\beta}_m \left(2^{k-1}t - n + 1\right), & \frac{n-1}{2^{k-1}} \leq t < \frac{n}{2^{k-1}}, \\ 0, & otherwise \end{cases} \tag{2.60}$$

with

$$\tilde{\beta}_m(t) = \begin{cases} 1, & m = 0, \\ \frac{1}{\sqrt{\frac{(-1)^{m-1}(m!)^2}{(2m)!}\alpha_{2m}}}\beta_m(t), & m > 0, \end{cases}$$

where $m = 0, 1, ..., M-1$ and $n = 1, 2, ..., 2^{k-1}$.

The coefficient $\frac{1}{\sqrt{\frac{(-1)^{m-1}(m!)^2}{(2m)!}\alpha_{2m}}}$ is for the orthonormality, the dilation parameter is $a = 2^{-(k-1)}$ and the translation parameter is $b = (n-1)2^{-(k-1)}$.

Here $\beta_m(t)$ are the well-known m^{th} order Bernoulli polynomials which are defined on the interval $[0, 1]$, and can be determined with the aid of the following explicit formula [96, 102]

$$\beta_m(t) = \sum_{i=0}^{m} \binom{m}{i} \alpha_{m-i}\, t^i,$$

where $\alpha_i,\quad i = 0, 1, ..., m$ are Bernoulli numbers .

The two dimensional Bernoulli wavelets can be defined as

$$\psi_{n_1,m_1,n_2,m_2}(t,r) = \begin{cases} 2^{\frac{k_1-1}{2}} 2^{\frac{k_2-1}{2}} \tilde{\beta}_{m_1}\left(2^{k_1-1}t - n_1 + 1\right) \tilde{\beta}_{m_2}\left(2^{k_2-1}r - n_2 + 1\right), \\ \qquad \frac{n_1-1}{2^{k_1-1}} \le t < \frac{n_1}{2^{k_1-1}},\ \frac{n_2-1}{2^{k_2-1}} \le r < \frac{n_2}{2^{k_2-1}}, \\ 0, \qquad otherwise, \end{cases} \tag{2.61}$$

where $m_1 = 0, 1, ..., M_1 - 1$, $m_2 = 0, 1, ..., M_2 - 1$, $n_1 = 1, 2, ..., 2^{k_1-1}$, and $n_2 = 1, 2, ..., 2^{k_2-1}$.

2.6.2 Properties of Bernoulli's polynomial

Properties of Bernoulli polynomials are given as follows [102]:

1. $\beta_m(1-t) = (-1)^m \beta_m(t), \qquad m \in Z^+$.
2. $\beta'_m(t) = m\beta_{m-1}(t), \qquad m \in Z^+$.
3. $\int_0^1 \beta_m(t)\beta_n(t)dt = (-1)^{m-1}\frac{m!n!}{(m+n)!}\alpha_{m+n}, \qquad m, n \ge 1$.
4. $\int_0^1 |\beta_m(t)|dt < 16\frac{m!}{(2\pi)^{m+1}}, \qquad m \ge 0$.
5. $\int_a^x \beta_m(t)dt = \frac{\beta_{m+1}(x) - \beta_{m+1}(a)}{m+1}$.
6. $\sup_{t\in[0,1]} |\beta_{2m}(t)| = |\alpha_{2m}|$.
7. $\sup_{t\in[0,1]} |\beta_{2m+1}(t)| \le \frac{2m+1}{4}|\alpha_{2m}|$.

2.6.3 Properties of Bernoulli number

The sequence of Bernoulli numbers $(\alpha_m)_{m\in N}$ satisfies the following properties [102]:

1. $\alpha_{2m+1} = 0$, $\alpha_{2m} = \beta_{2m}(1)$.
2. $\beta_m(1/2) = (2^{1-m} - 1)\alpha_m$.
3. $\alpha_m = -\frac{1}{m+1}\sum_{k=0}^{m-1}\binom{m+1}{k}\alpha_k$.

2.6.4 Function approximation

A function $f(x)$ defined over $[0,1)$ can be approximated by Bernoulli wavelets as

$$f(x) = \sum_{n=1}^{2^{k-1}}\sum_{m=0}^{M-1} c_{n,m}\psi_{n,m}(x) = C^T\Psi(x), \tag{2.62}$$

where C and $\Psi(x)$ are $\left(2^{k-1}M \times 1\right)$ column vectors given by

$$C = \left[c_{1,0}, c_{1,1}, ..., c_{1,M-1}, c_{2,0}, ..., c_{2,M-1}, ..., c_{2^{k-1},0}, ..., c_{2^{k-1},M-1}\right]^T, \tag{2.63}$$

$$\Psi(x) = \left[\psi_{1,0}(x), \psi_{1,1}(x), ..., \psi_{1,M-1}(x), ..., \psi_{2^{k-1},0}(x), ..., \psi_{2^{k-1},M-1}(x)\right]^T. \tag{2.64}$$

The function $f(x,t) \in [0,1) \times [0,1)$ can be approximated by two dimensional Bernoulli wavelets as

$$f(x,t) = \sum_{n_1=1}^{2^{k_1-1}}\sum_{m_1=0}^{M_1-1}\sum_{n_2=1}^{2^{k_2-1}}\sum_{m_2=0}^{M_2-1} c_{n_1,m_1,n_2,m_2}\psi_{n_1,m_1,n_2,m_2}(x,t) = C^T\Psi(x,t), \tag{2.65}$$

where C and $\Psi(x,t)$ are $\left(2^{k_1-1}2^{k_2-1}M_1M_2 \times 1\right)$ matrices given by

$$C = \left[c_{1,0,1,0}, ..., c_{1,0,1,M_2-1}, ..., c_{1,0,2^{k_2-1},M_2-1}, ..., c_{2^{k_1-1},M_1-1,2^{k_2-1},M_2-1}\right]^T, \tag{2.66}$$

$$\begin{aligned}\Psi(x,t) = [\psi_{1,0,1,0}(x,t), ..., \psi_{1,0,1,M_2-1}(x,t), ..., \psi_{1,0,2^{k_2-1},M_2-1}(x,t), ...\\ ..., \psi_{2^{k_1-1},M_1-1,2^{k_2-1},M_2-1}(x,t)]^T.\end{aligned} \tag{2.67}$$

2.6.5 Convergence analysis

Theorem 2.6.1. *If $f(x) \in L^2(\mathbb{R})$ be a continuous function defined on $[0,1]$ and $|f(x)| \le \tilde{M}$, then the Bernoulli wavelets expansion of $f(x)$ defined in eq. (2.62) converges uniformly and also*

$$|c_{n,m}| < \tilde{M}\frac{A}{2^{\frac{k-1}{2}}}\frac{16m!}{(2\pi)^{m+1}},$$

where $A = \frac{1}{\sqrt{\frac{(-1)^{m-1}(m!)^2}{(2m)!}\alpha_{2m}}}$.

Proof. Any function $f(x) \in L^2[0,1]$ can be expressed by Bernoulli wavelets as

$$f(x) = \sum_{n=1}^{2^{k-1}} \sum_{m=0}^{M-1} c_{n,m}\psi_{n,m}(x),$$

where the coefficients $c_{n,m}$ can be determined as

$$c_{n,m} = \langle f(x), \psi_{n,m}(x) \rangle .$$

Now for $n > 0$, $m > 0$,

$$\begin{aligned} c_{n,m} &= \langle f(x), \psi_{n,m}(x) \rangle \\ &= \int_0^1 f(x)\psi_{n,m}(x)dx \\ &= \int_{I_{nk}} f(x)\psi_{n,m}(x)dx \\ &= 2^{\frac{k-1}{2}} A \int_{I_{nk}} f(x)\beta_m(2^{k-1}x - n + 1)dx, \end{aligned}$$

where $I_{nk} = [\frac{n-1}{2^{k-1}}, \frac{n}{2^{k-1}})$.
Now, changing the variable $2^{k-1}x - n + 1 = t$, we have

$$c_{n,m} = \frac{1}{2^{\frac{k-1}{2}}} A \int_0^1 f\left(\frac{t+n-1}{2^{k-1}}\right) \beta_m(t)dt.$$

Thus,

$$\begin{aligned} |c_{n,m}| &\leq \frac{A}{2^{\frac{k-1}{2}}} \int_0^1 \left| f\left(\frac{t+n-1}{2^{k-1}}\right)\right| |\beta_m(t)|dt \\ &\leq \frac{A}{2^{\frac{k-1}{2}}} \tilde{M} \int_0^1 |\beta_m(t)|dt \\ &< \frac{A}{2^{\frac{k-1}{2}}} \tilde{M} \frac{16m!}{(2\pi)^{m+1}}, \text{ using the property of Bernoulli polynomials.} \end{aligned}$$

This means that the series $\sum_{n=1}^{2^{k-1}} \sum_{m=0}^{M-1} c_{n,m}$ is absolutely convergent and hence the series

$$\sum_{n=1}^{2^{k-1}} \sum_{m=0}^{M-1} c_{n,m}\psi_{n,m}(x)$$

is uniformly convergent [115]. □

2.7 Bernstein polynomial approximation

2.7.1 Bernstein polynomials and their properties

The general form of the Bernstein polynomials of n-th degree over the interval $[a, b]$ as defined in [33, 36, 183] is given by

$$B_{i,n}(x) = \binom{n}{i} \frac{(x-a)^i (b-x)^{n-i}}{(b-a)^n}, \qquad i = 0, 1, ..., n.$$

where $\binom{n}{i} = \frac{n!}{i!(n-i)!}$. Note that each of these $n+1$ polynomials having degree n satisfies the following properties:

$$\begin{aligned}
&B_{i,n}(x) = 0, \quad if \quad i < 0 \quad or \quad i > n, \\
&B_{i,n}(a) = B_{i,n}(b) = 0, \quad for \quad 1 \leq i \leq n-1, \\
&\sum_{i=0}^{n} B_{i,n}(x) = 1.
\end{aligned}$$

For $n = 10$, the Bernstein polynomial basis functions over the interval $[0, 1]$ are given as follows and the graph of all their basis functions are shown in Figure 2.1.

$$\begin{aligned}
B_{0,10}(x) &= (1-x)^{10}, \\
B_{1,10}(x) &= 10(1-x)^9 x, \\
B_{2,10}(x) &= 45(1-x)^8 x^2, \\
B_{3,10}(x) &= 120(1-x)^7 x^3, \\
B_{4,10}(x) &= 210(1-x)^6 x^4, \\
B_{5,10}(x) &= 252(1-x)^5 x^5, \\
B_{6,10}(x) &= 210(1-x)^4 x^6, \\
B_{7,10}(x) &= 120(1-x)^3 x^7, \\
B_{8,10}(x) &= 45(1-x)^2 x^8, \\
B_{9,10}(x) &= 10(1-x) x^9, \\
B_{10,10}(x) &= x^{10}.
\end{aligned}$$

2.7.2 Function approximation

Bernstein polynomials defined above form a complete basis [183] over the interval $[a, b]$. It is easy to show that any given polynomials of degree n can

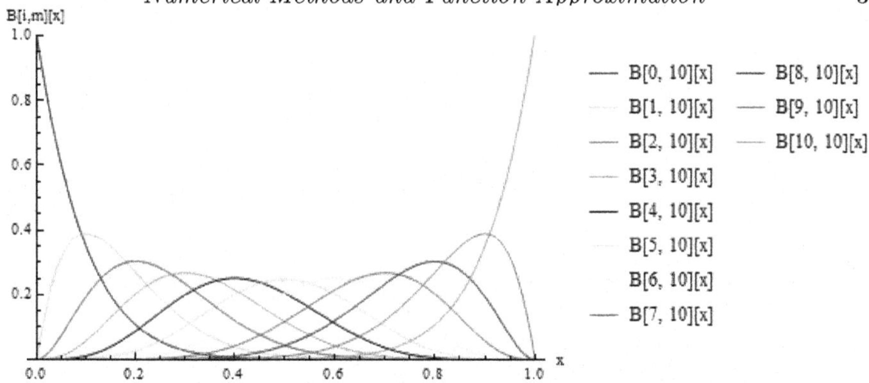

FIGURE 2.1: Graph of 10-degree Bernstein polynomials over $[0, 1]$.

be expressed in terms of linear combination of the basis functions. A function $y(x)$ defined over $[a, b]$ can be approximated by Bernstein polynomials basis functions of degree n as

$$y(x) \approx \sum_{i=0}^{n} c_i B_{i,n}(x) = C^T B(x), \tag{2.68}$$

where C and $B(x)$ are $(n+1) \times 1$ vectors defined as

$$C = [c_0, c_1, ..., c_n]^T,$$

$$B(x) = [B_{0,n}(x), B_{1,n}(x), ..., B_{n,n}(x)]^T.$$

2.8 Hybrid Legendre Block-Pulse functions

2.8.1 Legendre polynomials

Consider the Legendre polynomials $L_m(x)$ on the interval $[-1, 1]$ as

$$\begin{aligned} L_0(x) &= 1, \\ L_1(x) &= x, \\ L_m(x) &= \frac{2m-1}{m} x L_{m-1}(x) - \frac{m-1}{m} L_{m-2}(x), \quad m = 2, 3, ... \end{aligned}$$

The set $L_m(x) : m = 0, 1, 2, ...$ in Hilbert space $L^2[-1, 1]$ is a complete orthogonal set.

2.8.2 Block-Pulse functions

A set of Block-Pulse functions $b_n(x),\ \ n = 1, 2, ..., N$ on the interval $[0, 1)$ are defined as follows:

$$b_n(x) = \begin{cases} 1, & \frac{n-1}{N} \leq x < \frac{n}{N}, \\ 0, & otherwise. \end{cases}$$

The Block-Pulse functions on interval $[0, 1)$ are disjoint and also these functions are orthogonal on the interval $[0, 1)$.

2.8.3 Hybrid Legendre Block-Pulse functions

For $m = 0, 1, ..., M-1$ and $n = 1, 2, ..., N$, the hybrid Legendre Block-Pulse functions defined on the interval $[0, 1)$ are as follows [129, 124]:

$$b(n, m, x) = \begin{cases} \sqrt{N(2m+1)}L_m(2Nx - 2n + 1), & \frac{n-1}{N} \leq x < \frac{n}{N}, \\ 0, & otherwise, \end{cases} \tag{2.69}$$

where m and n are the order of the Legendre polynomials and Block-Pulse functions, respectively. The coefficient $\sqrt{N(2m+1)}$ is for the orthonormality.

2.8.4 Function approximation

Any function $f(x) \in L^2[0, 1)$ can be expressed by Hybrid Legendre-Block-Pulse functions as

$$f(x) = \sum_{n=1}^{\infty} \sum_{m=0}^{\infty} X(n, m)b(n, m, x). \tag{2.70}$$

If the above series is truncated at some values of M and N, then eq. (2.70) can be written as

$$f(x) \approx \sum_{n=1}^{N} \sum_{m=0}^{M-1} X(n, m)b(n, m, x) = X^T B(x), \tag{2.71}$$

where

$$X = [x(1,0), X(1,1), ..., X(1, M-1), ..., X(N, 0), X(N, 1), ..., X(N, M-1)]^T, \tag{2.72}$$

$$B(x) = [b(1, 0, x), ..., b(1, M-1, x), ..., b(N, 0, x), ..., b(N, M-1, x)]^T. \tag{2.73}$$

Since $\int_0^1 B(x)B^T(x)dx = I$, where I is an identity matrix of dimension NM, then we can calculate $X(n, m)$ as

$$X(n, m) = \int_0^1 f(x)b(n, m, x)dx, \qquad n = 1, 2, ..., N, \quad m = 0, 1, ..., M-1. \tag{2.74}$$

2.9 Sinc basis functions

2.9.1 Properties and approximation of Sinc functions

The Sinc function [160, 226] is defined on the real line, $-\infty < x < \infty$ by

$$Sinc(x) = \begin{cases} \frac{\sin(\pi x)}{\pi x}, & x \neq 0, \\ 1, & x = 0. \end{cases}$$

For $h > 0$, the translated Sinc functions with evenly spaced nodes are given by

$$S(j,h)(x) = Sinc\left(\frac{x - jh}{h}\right), \quad j = 0, \pm 1, \pm 2, ...$$

The Sinc function for the interpolating points $x_k = kh$ is given by

$$S(j,h)(kh) = \delta_{jk}^{(0)} = \begin{cases} 1, & k = j, \\ 0, & k \neq j. \end{cases}$$

Let D be a simply connected domain, having boundary ∂D. Let a and b denote two distinct points of ∂D and denote a conformal map of D_d onto D, where D_d denote the region $\{w \in C : \ |Im \ \ (t)| < d\}$ such that $\varphi(-\infty) = a$ and $\varphi(\infty) = b$. Let $\varphi^{-1}(D_d)$ denote the inverse map. Given φ and a positive number h, let us set $x_k = \varphi(kh)$, $k = 0, \pm 1, \pm 2, ...$

Let $H^1(D_d)$ the family of all functions f analytic in D_d, such that if $D_d(\varepsilon)$ is defined for $0 < \varepsilon < 1$ by

$$D_d(\varepsilon) = \left\{x \in C : \quad |Re \ x| < \frac{1}{\varepsilon}, \quad |Im \ x| < d(1 - \varepsilon)\right\}.$$

A function f is said to decay double exponentially with respect to the conformal map φ if there exist positive constants α and C such that

$$|f(\varphi(x))\varphi'(x)| \leq C \exp(-\alpha \exp|x|), \quad x \in (-\infty, \infty). \tag{2.75}$$

Let a be a positive number, then the space $K_\varphi^a(D_d)$ denotes the family of function f where $f(\varphi(x))\varphi'(x) \in H^1(D_d)$ and the analytic function f satisfies eq. (2.75).

To construct an approximation on the interval $[a, b]$, we consider the conformal map [226]

$$\varphi(x) = \frac{b - a}{2} \tanh(\frac{\pi}{2} \sinh x) + \frac{b + a}{2},$$

$$\varphi'(x) = \frac{b - a}{2} \frac{\frac{\pi}{2} \cosh x}{\cosh^2(\frac{\pi}{2} \sinh x)}.$$

Let N be a positive integer, and let mesh size h be selected by the formula

$$h = \frac{1}{N} \log\left(\frac{2\pi d N}{\alpha}\right);$$

then, we have the double exponential formula [226] for the definite integration of a function $y \in K_{\varphi}^{a}(D_d)$ as

$$\int_a^b y(x)dx = h \sum_{k=-N}^{N} y(\varphi(kh))\varphi'(kh) + O\left(\exp\left(-\frac{2\pi d N}{\log(2\pi d N/\alpha)}\right)\right). \quad (2.76)$$

Chapter 3

Numerical Solutions of Fredholm Integral Equations by B-spline Wavelet Method

3.1 Introduction

A computational approach to solve integral equations is an essential work in scientific research. Integral equations have been one of the essential tools for various areas of applied mathematics. Integral equations occur naturally in many fields of science and engineering [208]. Wavelets theory is a relatively new and emerging area in mathematical research. It has been applied in a wide range of engineering disciplines; particularly, wavelets are very successfully used in signal analysis for waveform representations and segmentations, time frequency analysis, and fast algorithms for easy implementation [53, 54]. Wavelets permit the accurate representation of a variety of functions and operators. Moreover, wavelets establish a connection with fast numerical algorithms. Wavelets can be separated into two distinct types, orthogonal and semi-orthogonal [53, 68]. The research works available in open literature on integral equations have shown a marked preference for orthogonal wavelets [148]. This is probably because the original wavelets, which were widely used for signal processing, were primarily orthogonal. In signal processing applications, unlike integral equation methods, the wavelet itself is never constructed since only its scaling function and coefficients are needed. However, orthogonal wavelets either have infinite support or a non-symmetric, and in some cases fractal, nature. These properties can make them a poor choice for characterization of a function. In contrast, the semi-orthogonal wavelets have finite support, both even and odd symmetry, and simple analytical expressions, ideal attributes of a basis function [148].

In this chapter, we apply compactly supported linear semi-orthogonal B-spline wavelets, specially constructed for the bounded interval to approximate the unknown function present in the integral equations. Semi-orthogonal wavelets using B-spline specially constructed for the bounded interval and these wavelets can be represented in a closed form. This provides a compact support. Semi-orthogonal wavelets form the basis in the space $L^2(\mathbb{R})$. Using this basis, an arbitrary function in $L^2(\mathbb{R})$ can be expressed as the wavelet series. For the finite interval $[0, 1]$, the wavelet series cannot be completely presented by using this basis. This is because supports of some basis are truncated at the left or right end points of the interval. Hence a special basis has to be introduced into the wavelet expansion on the finite interval. These functions are referred to as the boundary scaling functions and boundary wavelet functions. The B-spline wavelet method has been applied to solve linear and nonlinear integral equations and their systems. The above method reduces the integral equations to systems of algebraic equations and then these systems can be solved by any usual numerical methods. Here, we have applied Newton's method with an appropriate initial guess for solving these systems.

A computational approach to solve integral equations is an essential work in scientific research. Some methods for solving the second kind Fredholm integral equation are available in open literature. In literature, the Petrov-Galerkin method and the iterated Petrov-Galerkin method [52, 91] have been applied to solve nonlinear integral equations. A variation of the Nystrom method for nonlinear integral equations of second kind was presented by Lardy [110]. The learned researchers Maleknejad et al. [132] proposed a numerical method for solving nonlinear Fredholm integral equations of the second kind using the Sinc-collocation method. In [113, 130], the Haar wavelet method and the rationalized Haar function method have been applied to solve a system of linear Fredholm integral equations. The authors, in [136, 126], proposed some numerical methods for solving a system of linear Fredholm integral equations of the second kind using Block-Pulse functions and Taylor-Series expansion method. A novel mesh-less technique termed the Random Integral Quadrature (RIQ) method has been developed by Zou and Li [232]. By applying this method, the governing equations in the integral form are discretized directly with the field nodes distributed randomly or uniformly, which is achieved by discretizing the integral governing equations with the generalized integral quadrature (GIQ) technique over a set of background virtual nodes, and then interpolating the function values at the virtual nodes over a set of field nodes with the Local Kriging method, where the field nodes are distributed either randomly or uniformly. The Fictitious Time Integration Method (FTIM), previously developed by Liu and Atluri [114], has been employed to solve a system of ill-posed linear algebraic equations, which may result from the discretization of a first-kind linear Fredholm integral equation. The Gauss-Legendre Nystrom method [95] has been applied for determining approximate solutions of Fredholm integral equations of the second kind on finite intervals. The authors recent continuous-kernel approach is generalized in order to accommodate kernels that are either singular or of limited continuous differentiability at a finite number of points within the interval of integration. Quadratic integral equations are a class of nonlinear integral equations having many important uses in engineering and sciences. The Adomian decomposition method has been applied to solve the quadratic integral equations of Volterra type [64].

In Section 3.2, the linear Fredholm integral equation of second kind has been solved by the B-spline wavelet method. In Section 3.3, we have applied B-spline wavelets for solving nonlinear Fredholm integral equations and then compared these results with the results obtained by the variational iterative method. The system of linear and nonlinear Fredholm integral equations of second kind has been solved by using the B-spline wavelet method in Section 3.4 and 3.5, respectively. Error analysis of the B-spline wavelet method has been discussed in Section 3.6 and Section 3.7 yields a concluding summary of the whole chapter.

3.2 Application of B-spline wavelet method for solving linear Fredholm integral equations of second kind

In this section, we apply compactly supported linear semi-orthogonal B-spline wavelets, specially constructed for the bounded interval to solve the second kind linear Fredholm integral equations of the form:

$$y(x) = f(x) + \int_0^1 K(x,t)y(t)dt, \quad 0 \le x \le 1, \tag{3.1}$$

where $K(x,t)$ and $f(x)$ are known functions and $y(x)$ is an unknown function to be determined.

In this section, linear Fredholm integral equation of the second kind of the form (3.1) has been solved by using B-spline wavelets. For this, we use eq. (2.13) of Chapter 2 to approximate $y(x)$ as

$$y(x) = C^T\Psi(x), \tag{3.2}$$

where $\Psi(x)$ is defined in eq. (2.15) of Chapter 2, and C is $(2^{M+1}+1)\times 1$ unknown vector defined similarly as in eq. (2.14) of Chapter 2. We also expand $y(x)$ and $K(x,t)$ by B-spline dual wavelets $\tilde{\Psi}$ defined as in eqs. (2.22)-(2.23) of Chapter 2 as

$$f(x) = C_1^T\tilde{\Psi}(x), \quad K(x,t) = \tilde{\Psi}^T(t)\Theta\tilde{\Psi}(x), \tag{3.3}$$

where

$$\Theta_{i,j} = \int_0^1 \left[\int_0^1 K(x,t)\Psi_i(t)dt\right]\Psi_j(x)dx. \tag{3.4}$$

From eqs. (3.3) and (3.2), we get

$$\begin{aligned}\int_0^1 K(x,t)y(t)dt &= \int_0^1 C^T\Psi(t)\tilde{\Psi}^T(t)\Theta\tilde{\Psi}(x)dt \\ &= C^T\Theta\tilde{\Psi}(x),\end{aligned} \tag{3.5}$$

since

$$\int_0^1 \Psi(t)\tilde{\Psi}^T(t)dt = I.$$

By applying eqs. (3.2)-(3.5) in eq. (3.1), we have

$$C^T\Psi(x) - C^T\Theta\tilde{\Psi}(x) = C_1^T\tilde{\Psi}(x). \tag{3.6}$$

By multiplying both sides of the eq. (3.6) with $\Psi^T(x)$ from the right and integrating with respect to x from 0 to 1, we get

$$C^TP - C^T\Theta = C_1^T, \tag{3.7}$$

since

$$\int_0^1 \tilde{\Psi}(x)\Psi^T(x)dx = I,$$

and P is a $(2^{M+1}+1)\times(2^{M+1}+1)$ square matrix given by

$$P = \int_0^1 \Psi(x)\Psi^T(x)dx = \begin{bmatrix} P_1 & 0 \\ 0 & P_2 \end{bmatrix}. \tag{3.8}$$

Consequently, from equation (3.7), we get $C^T = C_1^T(P-\Theta)^{-1}$. Hence we can calculate the solution for $y(x) = C^T\Psi(x)$.

TABLE 3.1: Approximate solutions for $M = 2$ for Example 3.2.1

x	$y_{approximate}$	y_{exact}	Absolute error
0	1.001300	1.000000	1.30173E-3
0.1	0.995052	0.995004	4.75992E-5
0.2	0.979500	0.980067	5.66575E-4
0.3	0.954792	0.955336	5.44546E-4
0.4	0.921120	0.921061	5.94170E-5
0.5	0.878726	0.877583	1.14300E-3
0.6	0.825360	0.825336	2.45777E-5
0.7	0.764394	0.764842	4.47947E-4
0.8	0.696316	0.696707	3.90444E-4
0.9	0.621667	0.621610	5.68924E-5
1	0.541039	0.540302	7.36347E-4

TABLE 3.2: Approximate solutions for $M = 4$ for Example 3.2.1

x	$y_{approximate}$	y_{exact}	Absolute error
0	1.000080	1.000000	8.13789E-5
0.1	0.995007	0.995004	3.28342E-6
0.2	0.980032	0.980067	3.50527E-5
0.3	0.955302	0.955336	3.42873E-5
0.4	0.921064	0.921061	2.80525E-6
0.5	0.877654	0.877583	7.14185E-5
0.6	0.825339	0.825336	2.96120E-6
0.7	0.764815	0.764842	2.72328E-5
0.8	0.696682	0.696707	2.51241E-5
0.9	0.621612	0.621610	1.63566E-6
1	0.540347	0.540302	4.44686E-5

3.2.1 Illustrative examples

Example 3.2.1. *Consider the equation*

$$y(x) = \cos x + \frac{3}{2} x \sin x + \int_0^1 K(x,t) y(t) dt, \qquad 0 \leq x \leq 1,$$

where

$$K(x,t) = \begin{cases} -3\sin(x-t), & 0 \leq t \leq x \\ 0, & x \leq t \leq 1. \end{cases}$$

The solution $y(x)$ is obtained by B-spline wavelet method (BWM) for $M = 2, M = 4$. The numerical approximate results for $M = 2, M = 4$ together with their exact solutions $y(x) = \cos x$ and absolute errors are cited in Tables 3.1 and 3.2 respectively.

TABLE 3.3: Approximate solutions for $M = 2$ for Example 3.2.2

x	$y_{approximate}$	y_{exact}	Absolute error
0	-0.001751	0.000000	1.75070E-3
0.1	0.157913	0.157983	7.01720E-5
0.2	0.330182	0.329412	7.70007E-4
0.3	0.515056	0.514286	7.69838E-4
0.4	0.712534	0.712605	7.06777E-5
0.5	0.922618	0.924370	1.75154E-3
0.6	1.149510	1.149580	7.10762E-5
0.7	1.389000	1.388240	7.69042E-4
0.8	1.641100	1.640340	7.68812E-4
0.9	1.905810	1.905880	7.17658E-5
1	2.183120	2.184870	1.75269E-3

The error function is given by

Error function$= \|y_{exact}(x_i) - y_{approximate}(x_i)\|$

$$= \sqrt{\sum_{i=1}^{n}(y_{exact}(x_i) - y_{approximate}(x_i))^2}$$

Global error estimate=R.M.S.error

$$= \frac{1}{\sqrt{n}}\sqrt{\sum_{i=1}^{n}(y_{exact}(x_i) - y_{approximate}(x_i))^2}$$

In Example 3.2.1, *Error estimates (or R.M.S. errors)* are 0.00064165 and 0.0000398951 for $M = 2$ and $M = 4$ respectively.

Example 3.2.2. *Consider the equation*

$$y(x) = x + \int_0^1 (xt^2 + x^2t)y(t)dt, \quad 0 \le x \le 1.$$

The solution $y(x)$ is obtained by B-spline wavelet method (BWM) for $M = 2, M = 4$. The numerical approximate results for $M = 2, M = 4$ together with their exact solutions $y(x) = \frac{180x+80x^2}{119}$ and the absolute errors are cited in Tables 3.3 and 3.4, respectively.

In Example 3.2.2, *Error estimates (or R.M.S. errors)* are 0.0010266 and 0.0000641496 for $M = 2$ and $M = 4$ respectively.

TABLE 3.4: Approximate solutions for $M = 4$ for Example 3.2.2

x	$y_{approximate}$	y_{exact}	$Absoluteerror$
0	-0.000109	0.000000	1.09419E-4
0.1	0.157979	0.157983	4.37731E-6
0.2	0.329460	0.329412	4.81431E-5
0.3	0.514334	0.514286	4.81424E-5
0.4	0.712601	0.712605	4.37929E-6
0.5	0.924260	0.924370	1.09422E-4
0.6	1.149580	1.149580	4.38085E-6
0.7	1.388280	1.388240	4.81393E-5
0.8	1.640380	1.640340	4.81384E-5
0.9	1.905880	1.905880	4.38354E-6
1	2.184760	2.184870	1.09427E-4

TABLE 3.5: Approximate solutions for $M = 2$ for Example 3.2.3

x	$y_{approximate}$	y_{exact}	Absolute error
0	0.001187	0.000000	0.001187
0.1	0.310083	0.309017	0.001065
0.2	0.584716	0.587785	0.003069
0.3	0.804107	0.809017	0.004910
0.4	0.951212	0.951057	0.000155
0.5	1.012920	1.000000	0.012924
0.6	0.951212	0.951057	0.000155
0.7	0.804107	0.809017	0.004910
0.8	0.584716	0.587785	0.003069
0.9	0.310083	0.309017	0.001065
1	0.001187	0.000000	0.001187

Example 3.2.3. *Consider the equation*

$$y(x) = \left(1 - \frac{1}{\pi^2}\right)\sin(\pi x) + \int_0^1 K(x,t)y(t)dt, \quad 0 \le x \le 1,$$

where

$$K(x,t) = \begin{cases} x(1-t), & x \le t \\ t(1-x), & t \le x. \end{cases}$$

The solution $y(x)$ is obtained by B-spline wavelet method (BWM) for $M = 2, M = 4$. The numerical approximate results for $M = 2, M = 4$ together with their exact solutions $y(x) = \sin(\pi x)$ and absolute errors are cited in Tables 3.5 and 3.6, respectively.

In example 3.2.3, *error estimates (or R.M.S. errors)* are 0.00466338 and 0.000285911 for $M = 2$ and $M = 4$, respectively.

TABLE 3.6: Approximate solutions for $M = 4$ for Example 3.2.3

x	$y_{approximate}$	y_{exact}	Absolute error
0	1.823150E-5	0.000000	1.82315E-5
0.1	0.309012	0.309017	4.75611E-6
0.2	0.587571	0.587785	2.14100E-4
0.3	0.808735	0.809017	2.81802E-4
0.4	0.951092	0.951057	3.51844E-5
0.5	1.000800	1.000000	8.03434E-4
0.6	0.951092	0.951057	3.51844E-5
0.7	0.808735	0.809017	2.81802E-4
0.8	0.587571	0.587785	2.14100E-4
0.9	0.309012	0.309017	4.75611E-6
1	1.823150E-5	0.000000	1.82315E-5

3.3 Application of B-spline wavelet method for solving nonlinear Fredholm integral equations of second kind

In this section, we consider the second kind nonlinear Fredholm integral equation of the following form

$$u(x) = f(x) + \int_0^1 K(x,t)F(t,u(t))dt, \qquad 0 \leq x \leq 1. \tag{3.9}$$

where $K(x,t)$ is the kernel of the integral equation, $f(x)$ and $K(x,t)$ are known functions and $u(x)$ is the unknown function that is to be determined.
First, we assume

$$y(x) = F(x, u(x)), \qquad 0 \leq x \leq 1. \tag{3.10}$$

Now from eq. (2.13) of Chapter 2, we can approximate the functions $u(x)$ and $y(x)$ as

$$u(x) = A^T\Psi(x), \qquad y(x) = B^T\Psi(x), \tag{3.11}$$

where A and B are $(2^{M+1}+1) \times 1$ column vectors similar to C defined in eq. (2.14) of Chapter 2.
Again by using dual wavelet functions, we can approximate the functions $f(x)$ and $K(x,t)$ as follows

$$f(x) = D^T\tilde{\Psi}(x), \qquad K(x,t) = \tilde{\Psi}^T(t)\Theta\tilde{\Psi}(x), \tag{3.12}$$

where

$$\Theta_{(i,j)} = \int_0^1 \left[\int_0^1 K(x,t)\Psi_i(t)dt\right] \Psi_j(x)dx.$$

From eqs. (3.10)-(3.12), we get

$$\begin{aligned}\int_0^1 K(x,t)F(t,u(t))dt &= \int_0^1 B^T\Psi(t)\tilde{\Psi}^T(t)\Theta\tilde{\Psi}(x)dt \\ &= B^T\left[\int_0^1 \Psi(t)\tilde{\Psi}^T(t)dt\right]\Theta\tilde{\Psi}(x) \\ &= B^T\Theta\tilde{\Psi}(x), \end{aligned} \tag{3.13}$$

since $\int_0^1 \Psi(t)\tilde{\Psi}^T(t)dt = I$.
Applying eqs. (3.10)-(3.13) in eq. (3.9), we get

$$A^T\Psi(x) = D^T\tilde{\Psi}(x) + B^T\Theta\tilde{\Psi}(x). \tag{3.14}$$

Multiplying eq. (3.14) by $\psi^T(x)$ both sides from the right and integrating from 0 to 1, we have

$$\begin{aligned} A^TP &= D^T + B^T\Theta \\ A^TP - D^T - B^T\Theta &= 0, \end{aligned} \tag{3.15}$$

where P is a $(2^{M+1}+1)\times(2^{M+1}+1)$ square matrix given by

$$P = \int_0^1 \Psi(x)\Psi^T(x)dx = \begin{bmatrix} P_1 & \\ & P_2 \end{bmatrix}, and \quad \int_0^1 \tilde{\Psi}(x)\Psi^T(x)dx = I.$$

Eq. (3.15) gives a system of $(2^{M+1}+1)$ algebraic equations with $2(2^{M+1}+1)$ unknowns for A and B vectors given in eq. (3.11).
To find the solution $u(x)$ in eq. (3.11), we first utilize the following equation

$$F(x, A^T\Psi(x)) = B^T\Psi(x), \tag{3.16}$$

with the collocation points $x_i = \frac{i-1}{2^{M+1}}$, where $i = 1, 2, ..., 2^{M+1}+1$.
Eq. (3.16) gives a system of $(2^{M+1}+1)$ algebraic equations with $2(2^{M+1}+1)$ unknowns for A and B vectors given in eq. (3.11).

Combining eqs. (3.15) and (3.16), we have a total of $2(2^{M+1}+1)$ algebraic equations with the same number of unknowns for A and B. Solving those equations for the unknown coefficients in the vectors A and B, we can obtain the solution $u(x) = A^T\Psi(x)$. The obtained results have been compared with the results obtained by variational iteration method (VIM).

- **VIM technique for eq.** (3.9)

For solving eq. (3.9) by variational iteration method, first we have to take the partial derivative of eq. (3.9) with respect to x.

$$u'(x) = f'(x) + \int_0^1 \frac{\partial K(x,t)}{\partial x}F(t,u(t))dt. \tag{3.17}$$

We apply the variational iteration method for the eq. (3.17). According to this method, correction functional can be defined as

$$u_{n+1}(x) = u_n(x) + \int_0^x \lambda(\xi) \left(u_n'(\xi) - f'(\xi) - \int_a^b \frac{\partial K(\xi,t)}{\partial \xi} F(t, \tilde{u}_n(t))dt \right) d\xi, \tag{3.18}$$

where $\lambda(\xi)$ is a general Lagrange multiplier which can be identified optimally by the variational theory, the subscript n denotes the nth order approximation and $\tilde{u}_n$ is considered as a restricted variation, i.e., $\delta\tilde{u}_n = 0$. The successive approximations $u_n(x), n \geq 1$ for the solution $u(x)$ can be readily obtained after determining the Lagrange multiplier and selecting an appropriate initial function $u_0(x)$. Consequently, the approximate solution may be obtained by using $u(x) = \lim_{n\to\infty} u_n(x)$.

To make the above correction functional stationary, we have

$$\begin{aligned}
\delta u_{n+1}(x) &= \delta u_n(x) + \delta \int_0^x \lambda(\xi) \left(u_n'(\xi) - f'(\xi) - \int_a^b \frac{\partial K(\xi,t)}{\partial \xi} F(t, \tilde{u}_n(t))dt \right) d\xi \\
&= \delta u_n(x) + \int_0^x \lambda(\xi)\delta(u_n'(\xi))d\xi \\
&= \delta u_n(x) + \lambda \delta u_n|_{\xi=x} - \int_0^x \lambda'(\xi)\delta u_n(\xi)d\xi.
\end{aligned} \tag{3.19}$$

Under the stationary condition, $\delta u_{n+1} = 0$, implies the following Euler Lagrange equation

$$\lambda'(\xi) = 0, \tag{3.20}$$

with the following natural boundary condition

$$1 + \lambda(\xi)|_{\xi=x} = 0. \tag{3.21}$$

Solving eq. (3.20) along with boundary condition (3.21), we get the general Lagrange multiplier $\lambda = -1$.

Substituting the identified Lagrange multiplier into eq. (3.18), results in the following iterative scheme

$$u_{n+1}(x) = u_n(x) - \int_0^x \left(u_n'(\xi) - f'(\xi) - \int_a^b K'(\xi,t)F(t, \tilde{u}_n(t))dt \right) d\xi, \quad n \geq 0. \tag{3.22}$$

By starting with the initial approximate function $u_0(x) = f(x)$ (say), we can determine the approximate solution $u(x)$.

TABLE 3.7: Comparison of numerical results of Example 3.3.1 by BWM and VIM

x	Exact	BWM		Error in BWM		VIM	Error in VIM
		$M=2$	$M=4$	$M=2$	$M=4$		
0.0	0	-2.496E-4	-3.617E-6	0.0002495	3.61701E-6	0	0
0.1	0.001	0.00166976	0.00105608	0.0006695	0.0000560	0.000965618	0.0000343
0.2	0.008	0.0105762	0.00816602	0.0025762	0.0001660	0.00792412	0.0000758
0.3	0.027	0.0313409	0.0272625	0.0043409	0.0002625	0.0268755	0.0001245
0.4	0.064	0.068719	0.0642781	0.0047190	0.0002781	0.0638198	0.0001802
0.5	0.125	0.12735	0.125145	0.0023496	0.0001454	0.124757	0.0002430
0.6	0.216	0.223711	0.216493	0.0077114	0.0004932	0.215687	0.0003130
0.7	0.343	0.355417	0.343778	0.0124172	0.0007776	0.34261	0.0003901
0.8	0.512	0.527106	0.512931	0.0151063	0.0009313	0.511526	0.0004743
0.9	0.729	0.743534	0.729887	0.0145338	0.0008869	0.728434	0.0005656
1.0	1	1.00957	1.00058	0.0095707	0.0005807	0.999336	0.0006641

3.3.1 Illustrative examples

Example 3.3.1. *Consider the equation*

$$u(x) = -\frac{x}{9} - \frac{x^2}{8} + x^3 + \int_0^1 (x^2 t + x t^2) u^2(t)dt, \qquad 0 \le x \le 1,$$

with the exact solution $u(x) = x^3$. The approximate solution is obtained by the B-spline wavelet method for $M = 2$ and $M = 4$ and also by VIM. The following Table 3.7 cites the numerical solutions obtained by B-spline method and VIM accomplished with corresponding exact solutions and the absolute errors. Figures 3.1-3.2 and Figures 3.3-3.4 present the comparison graphically between the numerical solutions obtained by B-spline wavelet method with exact solutions and VIM solutions respectively.

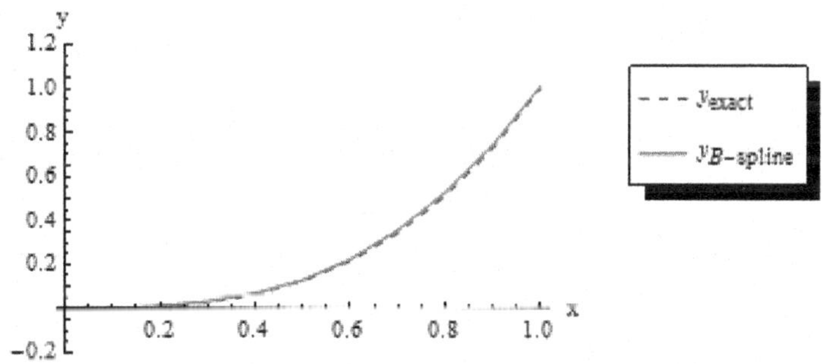

FIGURE 3.1: Comparison of numerical solution obtained by B-spline ($M = 2$) with exact solution for Example 3.3.1.

Example 3.3.2. *Consider the equation*

$$u(x) = \cos x - \frac{1}{2} x \sin(2) + \int_0^1 x(u^2(t) - \sin^2(t))dt, \qquad 0 \le x \le 1,$$

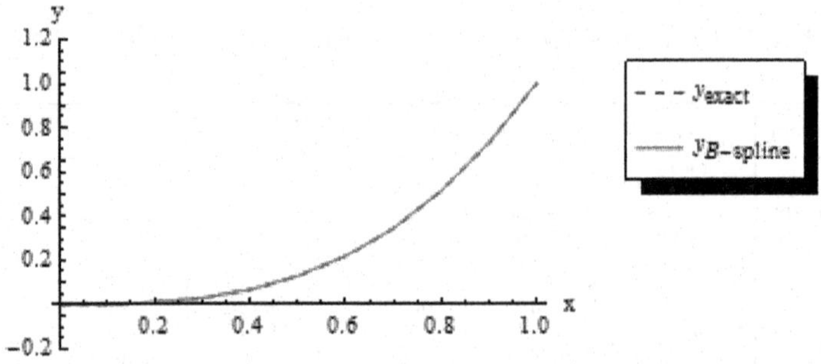

FIGURE 3.2: Comparison of numerical solution obtained by B-spline ($M = 4$) with exact solution for Example 3.3.1.

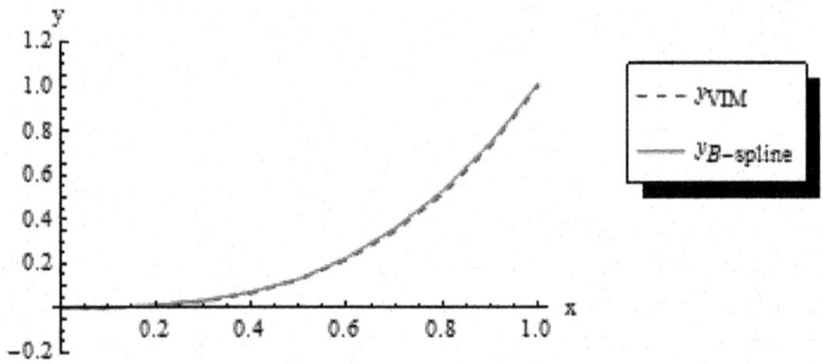

FIGURE 3.3: Comparison of numerical solution obtained by B-spline ($M = 2$) with VIM solution for Example 3.3.1.

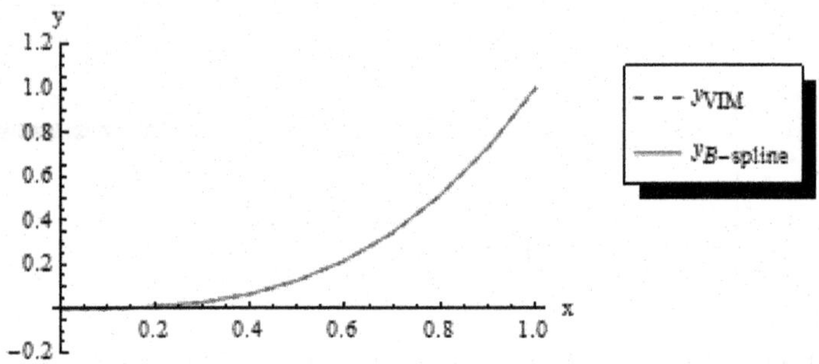

FIGURE 3.4: Comparison of numerical solution obtained by B-spline ($M = 4$) with VIM solution for Example 3.3.1.

with the exact solution $u(x) = \cos x$. *The approximate solution is obtained by the B-spline wavelet method for* $M = 2$ *and* $M = 4$ *and also by VIM. The following Table 3.8 cites the numerical solutions obtained by B-spline method and VIM accomplished with corresponding exact solutions and the absolute errors. Figures 3.5-3.6 and Figures 3.7-3.8 present the comparison graphically between the numerical solutions obtained by B-spline wavelet method with exact solutions and VIM solutions respectively.*

TABLE 3.8: Comparison of numerical results of Example 3.3.2 by BWM and VIM

x	Exact	BWM		Error in BWM		VIM	Error in VIM
		$M=2$	$M=4$	$M=2$	$M=4$		
0.0	1	1.0013	1.00008	0.0013027	0.0000813	1	0
0.1	0.995004	0.994854	0.994995	0.0001501	9.23203e-6	0.994897	0.0001069
0.2	0.980067	0.979103	0.980006	0.0009632	0.0000600	0.979853	0.0002139
0.3	0.955336	0.954196	0.955265	0.0011401	0.0000718	0.955016	0.0003209
0.4	0.921061	0.920326	0.921014	0.0007350	0.0000472	0.920633	0.0004279
0.5	0.877583	0.877732	0.877591	0.0001495	8.82354E-6	0.877048	0.0005348
0.6	0.825336	0.824168	0.825263	0.0011679	0.0000721	0.824694	0.0006418
0.7	0.764842	0.763003	0.764727	0.0018390	0.0001148	0.764093	0.0007488
0.8	0.696707	0.694726	0.696581	0.0019810	0.0001252	0.695851	0.0008558
0.9	0.62161	0.619877	0.621499	0.0017320	0.0001110	0.620647	0.0009628
1.0	0.540302	0.53905	0.540222	0.0012520	0.0000807	0.539233	0.0010697

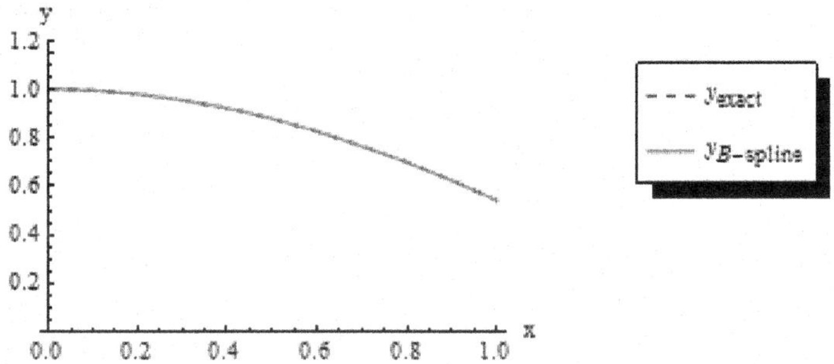

FIGURE 3.5: Comparison of numerical solution obtained by B-spline ($M=2$) with exact solution for Example 3.3.2.

Example 3.3.3. *Consider the equation*

$$u(x) = \frac{7}{8}x + \frac{1}{2}\int_0^1 xtu^2(t)dt, \qquad 0 \le x \le 1,$$

with the exact solution $u(x) = x$. The approximate solution is obtained by the B-spline wavelet method for $M = 2$ and $M = 4$ and also by VIM. The following Table 3.9 cites the numerical solutions obtained by B-spline method and VIM accomplished with corresponding exact solutions and the absolute errors. Figures 3.9-3.10 and Figures 3.11-3.12 present the comparison graphically between the numerical solutions obtained by B-spline wavelet method with exact solutions and VIM solutions, respectively.

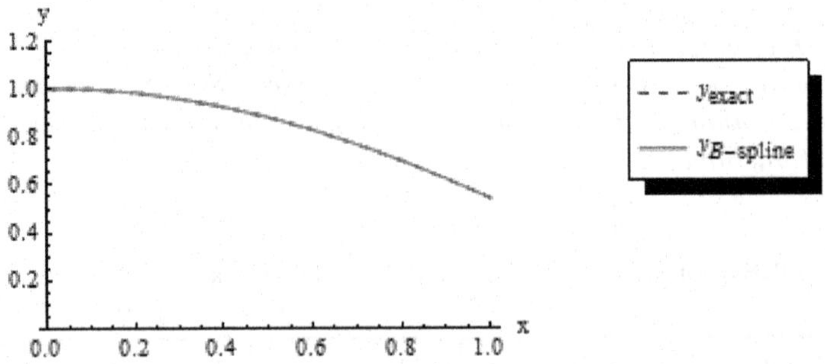

FIGURE 3.6: Comparison of numerical solution obtained by B-spline ($M = 4$) with exact solution for Example 3.3.2.

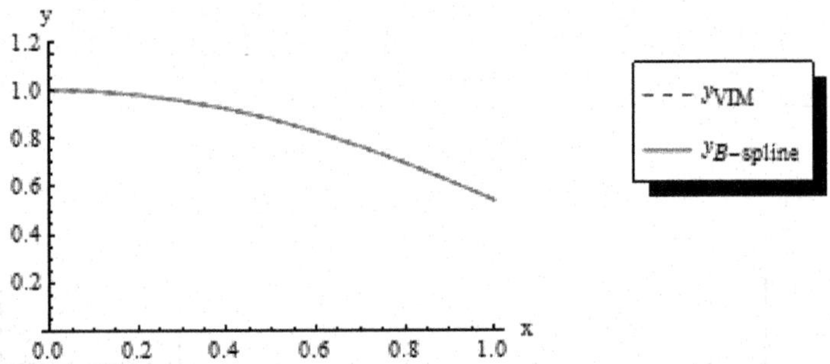

FIGURE 3.7: Comparison of numerical solution obtained by B-spline ($M = 2$) with VIM solution for Example 3.3.2.

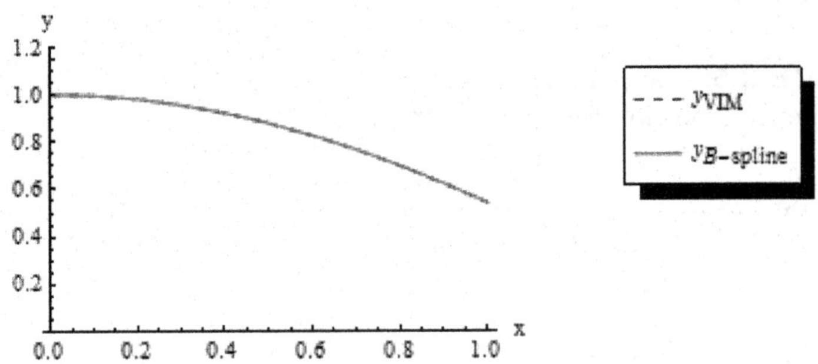

FIGURE 3.8: Comparison of numerical solution obtained by B-spline ($M = 4$) with VIM solution for Example 3.3.2.

TABLE 3.9: Comparison of numerical results of Example 3.3.3 by BWM and VIM

x	Exact	BWM		Error in BWM		VIM	Error in VIM
		$M=2$	$M=4$	$M=2$	$M=4$		
0.0	0.0	0	0	0	0	0	0
0.1	0.1	0.100087	0.100005	0.0000869	5.42599e-6	0.1	1.09593e-8
0.2	0.2	0.200174	0.200011	0.0001739	0.0000108	0.2	2.19186e-8
0.3	0.3	0.300261	0.300016	0.0002609	0.0000162	0.3	3.28778e-8
0.4	0.4	0.400348	0.400022	0.0003478	0.0000217	0.4	4.38371e-8
0.5	0.5	0.500435	0.500027	0.0004348	0.0000271	0.5	5.47964e-8
0.6	0.6	0.600522	0.600033	0.0005218	0.0000325	0.6	6.57557e-8
0.7	0.7	0.700609	0.700038	0.0006087	0.0000379	0.7	7.6715e-8
0.8	0.8	0.800696	0.800043	0.0006957	0.0000434	0.8	8.76743e-8
0.9	0.9	0.900783	0.900049	0.0007827	0.0000488	0.9	9.86335e-8
1.0	1.0	1.00087	1.00005	0.0008696	0.0000542	1.0	1.09593e-7

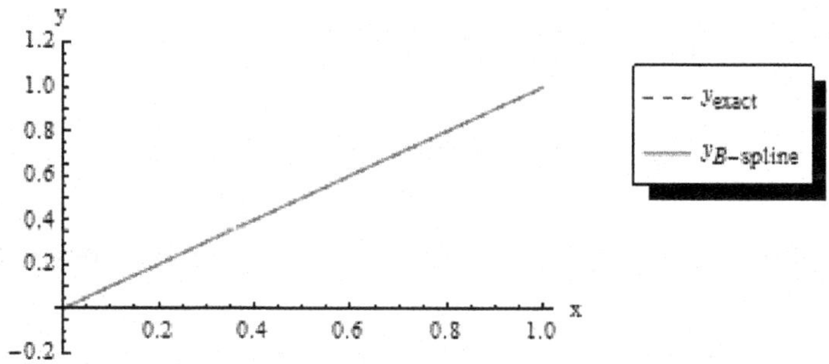

FIGURE 3.9: Comparison of numerical solution obtained by B-spline ($M = 2$) with exact solution for Example 3.3.3.

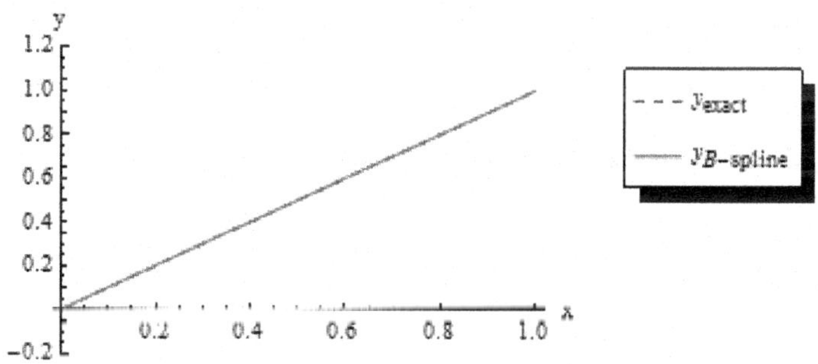

FIGURE 3.10: Comparison of numerical solution obtained by B-spline ($M = 4$) with exact solution for Example 3.3.3.

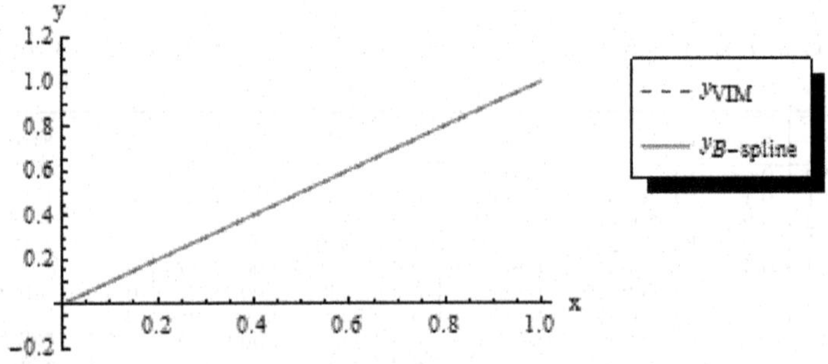

FIGURE 3.11: Comparison of numerical solution obtained by B-spline ($M = 2$) with VIM solution for Example 3.3.3.

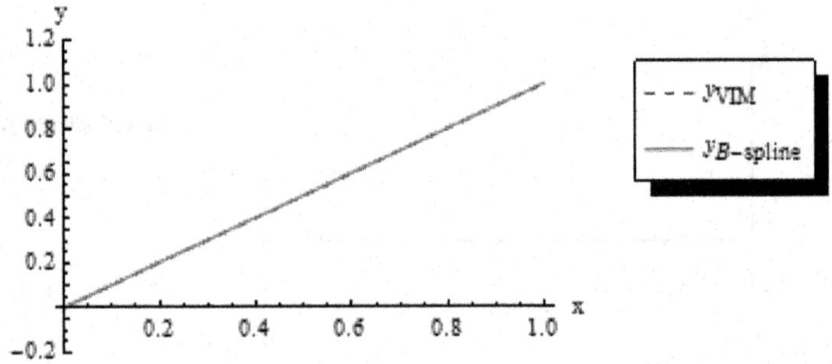

FIGURE 3.12: Comparison of numerical solution obtained by B-spline ($M = 4$) with VIM solution for Example 3.3.3.

Example 3.3.4. *Consider the equation*

$$u(x) = -\sin x - x^3\left(-\frac{367}{4096}\cos(4)\sin(4) + \frac{11357}{98304} - \frac{2095}{32768}\cos^2(4)\right) + \int_0^1 x^3t^5u^2(t)dt, \quad 0 \le x \le 1,$$

with the exact solution $u(x) = \sin(-4x)$. *The approximate solution is obtained by the B-spline wavelet method for* $M = 2$ *and* $M = 4$ *and also by VIM. The following Table 3.10 cites the numerical solutions obtained by B-spline method and VIM accomplished with corresponding exact solutions and the absolute errors. Figures 3.13-3.14 and Figures 3.15-3.16 present the comparison graphically between the numerical solutions obtained by B-spline wavelet method with exact solutions and VIM solutions, respectively.*

TABLE 3.10: Comparison of numerical results of Example 3.3.3 by BWM and VIM

x	Exact	BWM		Error in BWM		VIM	Error in VIM
		$M = 2$	$M = 4$	$M = 2$	$M = 4$		
0.0	0	-0.00247969	-0.00003766	0.0024796	0.0000376	0	0
0.1	-0.389418	-0.391555	-0.389409	0.0021369	9.45907e-6	-0.389418	3.53614e-8
0.2	-0.717356	-0.711074	-0.716931	0.0062817	0.0004255	-0.717356	2.82891e-7
0.3	-0.932039	-0.922784	-0.9315	0.0092549	0.0005389	-0.93204	9.54757e-7
0.4	-0.999574	-1.00002	-0.999602	0.0004443	0.0000283	-0.999576	2.26313e-6
0.5	-0.909297	-0.927696	-0.910438	0.0183982	0.0011404	-0.909302	4.42017e-6
0.6	-0.675463	-0.673206	-0.675445	0.0022571	0.0000183	-0.675471	7.63806e-6
0.7	-0.334988	-0.328869	-0.334689	0.0061188	0.0002991	-0.335	0.0000121
0.8	0.0583741	0.0599757	0.0585377	0.0016015	0.0001635	0.058356	0.0000181
0.9	0.44252	0.4454819	0.44283	0.0028986	0.0003096	0.442495	0.0000257
1.0	0.756809	0.77698	0.758118	0.0201778	0.0013154	0.756767	0.0000353

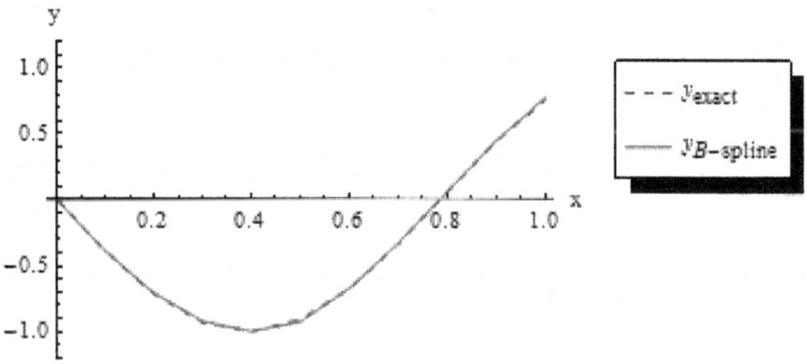

FIGURE 3.13: Comparison of numerical solution obtained by B-spline ($M = 2$) with exact solution for Example 3.3.4.

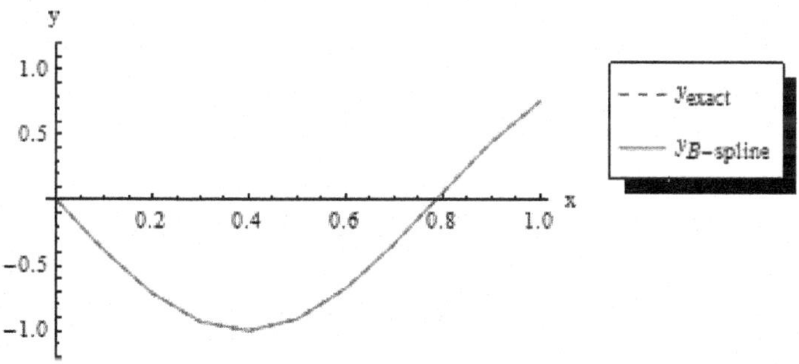

FIGURE 3.14: Comparison of numerical solution obtained by B-spline ($M = 4$) with exact solution for Example 3.3.4.

3.4 Application of B-spline wavelet method for solving system of linear Fredholm integral equations of second kind

In this section, we consider the system of linear Fredholm integral equations of second kind of the following form

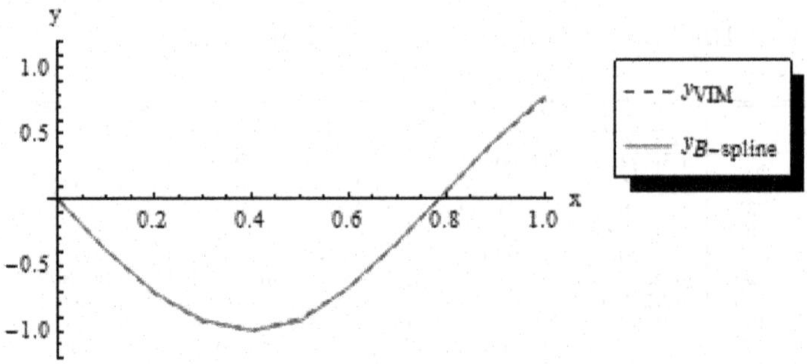

FIGURE 3.15: Comparison of numerical solution obtained by B-spline ($M = 2$) with VIM solution for Example 3.3.4.

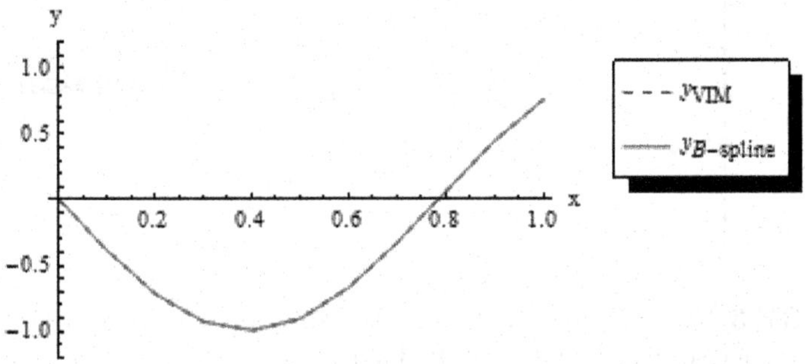

FIGURE 3.16: Comparison of numerical solution obtained by B-spline ($M = 4$) with VIM solution for Example 3.3.4.

$$\sum_{j=1}^{n} g_{i,j} y_j(x) = f_i(x) + \sum_{j=1}^{n} \int_0^1 K_{i,j}(x,t) y_j(t) dt, \quad i = 1, 2, ..., n, \tag{3.23}$$

where $f_i(x)$ and $K_{i,j}(x,t)$ are known functions and $y_j(x)$ are the unknown functions for $i, j = 1, 2, ..., n, \quad n \in \mathbb{N}$.

Now from eq. (2.13) of Chapter 2, we can approximate the following functions as

$$y_j(x) = A_j^T \Psi(x), \tag{3.24}$$

$$f_i(x) = B_i^T \Psi(x), \tag{3.25}$$

$$K_{i,j}(x,t) = \Psi^T(t) \Theta_{i,j} \Psi(x). \tag{3.26}$$

where A_j and B_j are $(2^{M+1}+1)\times 1$ column vectors similar to C defined in eq. (2.14) of Chapter 2 and

$$\Theta_{i,j} = \int_0^1 \left[\int_0^1 K_{i,j}(x,t)\tilde{\Psi}(t)dt\right]\tilde{\Psi}^T(x)dx.$$

We can calculate B_i^T as

$$B_i^T = \int_0^1 f_i(x)\tilde{\Psi}^T(x)dx.$$

From eqs. (3.24)-(3.26), we get

$$\begin{aligned}\int_0^1 K_{i,j}(x,t)y_j(t)dt &= \int_0^1 A_j^T\Psi(t)\Psi^T(t)\Theta_{i,j}\Psi(x)dt\\ &= A_j^T\left[\int_0^1 \Psi(t)\Psi^T(t)dt\right]\Theta_{i,j}\Psi(x)\\ &= A_j^T P\Theta_{i,j}\Psi(x),\end{aligned} \tag{3.27}$$

where P is a $(2^{M+1}+1)\times(2^{M+1}+1)$ square matrix given by

$$P = \int_0^1 \Psi(t)\Psi^T(t)dt = \begin{bmatrix} P_1 & \\ & P_2\end{bmatrix}.$$

Applying eqs. 3.24-3.27 in eq. 3.23, we get

$$\sum_{j=1}^{n} g_{i,j}A_j^T\Psi(x) = B_i^T\Psi(x) + \sum_{j=1}^{n} A_j^T P\Theta_{i,j}\Psi(x), \quad i = 1,2,...,n. \tag{3.28}$$

We can utilize the eq. (3.28) with the collocation points $x_s = \frac{s-1}{2^{M+1}}, \quad s = 1,2,...,2^{M+1}+1$.
Equation (3.28) reduces to a system of $n(2^{M+1}+1)$ linear equations with the same number of unknowns, which can be easily solved by Newton's method and thus we can obtain the solutions $y_j(x) = A_j^T\Psi(x)$, $j = 1,2,...,n$.

3.4.1 Algorithm

Input:

Read M, l, m, n, j_0, a, b (where the symbols have their usual meanings).
The cardinal B-spline functions of order $m \geq 2$ are

$$B_{1,j,X}(x) = \begin{cases} 1 & , x = [x_j, x_{j+1}) \\ 0 & , \textit{otherwise}\end{cases}$$

$$B_{m,j,X}(x) = \frac{x-x_j}{x_{j+m-1}-x_j}B_{m-1,j,X}(x) + \frac{x_{j+m}-x}{x_{j+m}-x_{j+1}}B_{m-1,j+1,X}(x),$$
$$j = -m+1,...,n-1$$

Knot sequence $X^{(j)} = \begin{cases} x^{(j)}_{-m+1} = ... = x^{(j)}_0 = 0, & \\ x^{(j)}_k = \frac{k}{2^j}, & k = 1, ..., n-1, \\ x^{(j)}_n = ... = x^{(j)}_{n+m-1} = 1. & \end{cases}$ and

$Supp\, B_{m,j,X}(x) = [x_j, x_{j+m}] \bigcap [a, b]$.

Initial step:

Construct scaling functions

$$\varphi_{m,j,i}(x) = \begin{cases} B_{m,j_0,i}(2^{j-j_0}x), & i = -m+1, ..., -1, \\ B_{m,j_0,2^j-m-i}(1 - 2^{j-j_0}x), & i = 2^j - m + 1, ..., 2^j - 1, \\ B_{m,j_0,0}(2^{j-j_0}x - 2^{j_0}i), & i = 0, ..., 2^j - m. \end{cases}$$

and wavelet functions

$$\psi_{m,j,i-m} = \sum_{k=i}^{2i+2m-2} q_{i,k} B_{m,j,k-m}, \ i = 1, ..., m-1,$$

$$\psi_{m,j,i-m} = \sum_{k=2i-m}^{2i+2m-2} q_{i,k} B_{m,j,k-m}, \ i = m, ..., n-m+1,$$

$$\psi_{m,j,i-m} = \sum_{k=2i-m}^{n+i+m-1} q_{i,k} B_{m,j,k-m}, \ i = n-m+2, ..., n.$$

Step 1:

Create

$$\Psi = [\varphi_{2,-1}, \varphi_{2,0}, ..., \varphi_{2,3}, \psi_{2,-1}, ..., \psi_{2,2}, \psi_{3,-1}, ..., \psi_{3,6}, ..., \psi_{M,-1}, ..., \psi_{M,2^M-2}]^T,$$

and

$$A_j = [c_{-1}, c_0, ..., c_3, d_{2,-1}, ..., d_{2,2}, d_{3,-1}, ..., d_{3,6}, ..., d_{M,-1}, ..., d_{M,2^M-2}]^T,$$

Next, compute $P = \int_0^1 \Psi(t)\Psi^T(t)dt$, $\tilde{\Psi} = P^{-1}\Psi$, $B_i = \int_0^1 f_i(x)\tilde{\Psi}^T(x)dx$, and $\Theta_{i,j} = \int_0^1 \left[\int_0^1 K_{i,j}(x,t)\tilde{\Psi}(t)dt\right] \tilde{\Psi}^T(x)dx$. **Step 2:**

Substitute $y_j(x) = A_j^T\Psi(x)$, $f_i(x) = B_i^T\Psi(x)$ and $K_{i,j}(x,t) = \Psi^T(t)\Theta_{i,j}\Psi(x)$ in eq. (3.23).

Then compute

$$\sum_0^l g_{i,j} A_j^T \Psi(x_s) = B_i^T \Psi(x_s) + \sum_{j=1}^l A_j^T P \Theta_{i,j} \Psi(x_s), \quad i = 1, 2, ..., l \tag{3.29}$$

using $x_s = \frac{s-1}{2^{M+1}}$, $s = 1, 2, ..., 2^{M+1} + 1$.

Eq. (3.29) leads to a system of $l(2^{M+1} + 1)$ linear equations with $l(2^{M+1} + 1)$ unknowns in A_j^T.

TABLE 3.11: Approximate solutions obtained by B-spline wavelet method and adaptive method based on trapezoidal rule along with exact solutions for Example 3.4.1

x	B-spline wavelet method				Adaptive method		Exact	
	$M=2$		$M=4$					
	$y_1(x)$	$y_2(x)$	$y_1(x)$	$y_2(x)$	$y_1(x)$	$y_2(x)$	$y_1(x)$	$y_2(x)$
0.0	2.21E-6	-0.0026	8.659E-9	-0.000163	-0.005342	0	0	0
0.1	0.100002	0.009896	0.1	0.009993	0.949896	0.009829	0.1	0.01
0.2	0.200002	0.041146	0.2	0.0400716	0.195322	0.039627	0.2	0.04
0.3	0.300002	0.091145	0.3	0.090016	0.295654	0.0893931	0.3	0.09
0.4	0.400002	0.159896	0.4	0.159993	0.395986	0.159127	0.4	0.16
0.5	0.500002	0.247396	0.5	0.249837	0.496318	0.24883	0.5	0.25
0.6	0.600001	0.359896	0.6	0.359993	0.59665	0.358501	0.6	0.36
0.7	0.700001	0.491146	0.7	0.490072	0.696982	0.488141	0.7	0.49
0.8	0.800001	0.641146	0.8	0.640072	0.797314	0.637749	0.8	0.64
0.9	0.900001	0.809896	0.9	0.809993	0.897646	0.807325	0.9	0.81
1.0	1	0.997396	1	0.999837	0.997978	0.99687	1	1

Step 3:
Solve eq. (3.29) by any numerical methods to get the solution for unknowns in A_j^T.
Output:
Solution for unknowns A_j^T.
Finally, compute $y_j(x) = A_j^T \Psi(x)$.
Stop

3.4.2 Illustrative examples

Example 3.4.1. *Consider the following system of Fredholm integral equations [136]*

$$y_1(x) = \frac{11}{6}x + \frac{11}{15} - \int_0^1 (x+t)y_1(t)dt - \int_0^1 (x+2t^2)y_2(t)dt,$$
$$y_2(x) = \frac{5}{4}x^2 + \frac{1}{4}x - \int_0^1 xt^2 y_1(t)dt - \int_0^1 x^2(t)y_2(t)dt,$$

with the exact solution $y_1(x) = x$ and $y_2(x) = x^2$. The comparison between the approximate solutions obtained by B-spline wavelet method for $M = 2$ and $M = 4$ and adaptive method based on trapezoidal rule [147] along with their exact solutions have been shown in Table 3.11. Table 3.12 cites the absolute errors obtained by these two methods.

Example 3.4.2. *Consider the following system of Fredholm integral equations [24]*

TABLE 3.12: Absolute errors obtained by B-spline wavelet method and adaptive method based on trapezoidal rule for Example 3.4.1

x	B-spline wavelet method				Adaptive method	
	$M=2$		$M=4$			
	$y_1(x)$	$y_2(x)$	$y_1(x)$	$y_2(x)$	$y_1(x)$	$y_2(x)$
0.0	2.2167E-6	0.00260417	8.65897E-9	0.00016276	0.00534236	0
0.1	2.07979E-6	0.000104205	8.12417E-9	6.51057E-6	0.00501037	0.000170677
0.2	1.94288E-6	0.00114576	7.58937E-9	7.16143E-5	0.00467838	0.000372991
0.3	1.80597E-6	0.00114572	7.05456E-9	7.16142E-5	0.00434638	0.000606941
0.4	1.66906E-6	0.000104308	6.51976E-9	6.51097E-6	0.00401439	0.000872527
0.5	1.53215E-6	0.00260434	5.98496E-9	1.62761E-4	0.0036824	0.00116975
0.6	1.39524E-6	0.000104367	5.45016E-9	6.5112E-6	0.0033504	0.00149861
0.7	1.25833E-6	0.00114561	4.91536E-9	7.16137E-5	0.00301841	0.0018591
0.8	1.12142E-6	0.00114558	4.38055E-9	7.16136E-5	0.00268642	0.00225123
0.9	9.84513E-7	0.000104438	3.84575E-9	6.51148E-6	0.00235442	0.002675
1.0	8.47603E-7	0.00260446	3.31095E-9	1.62762E-4	0.00202243	0.00313041

TABLE 3.13: Approximate solutions obtained by B-spline wavelet method and adaptive method based on trapezoidal rule along with exact solutions for Example 3.4.2

x	B-spline wavelet method				Adaptive method		Exact	
	$M=2$		$M=4$					
	$y_1(x)$	$y_2(x)$	$y_1(x)$	$y_2(x)$	$y_1(x)$	$y_2(x)$	$y_1(x)$	$y_2(x)$
0.0	1	0.997396	1	0.999837	1.00378	1	1	1
0.1	1.1	1.0099	1.1	1.00999	1.10423	1.01114	1.1	1.01
0.2	1.2	1.04115	1.2	1.04007	1.20467	1.04227	1.2	1.04
0.3	1.3	1.09115	1.3	1.09007	1.30512	1.09341	1.3	1.09
0.4	1.4	1.1599	1.4	1.15999	1.40556	1.16454	1.4	1.16
0.5	1.5	1.2474	1.5	1.24984	1.50601	1.25268	1.5	1.25
0.6	1.6	1.3599	1.6	1.35999	1.60646	1.36681	1.6	1.36
0.7	1.7	1.49115	1.7	1.49007	1.7069	1.49795	1.7	1.49
0.8	1.8	1.64115	1.8	1.64007	1.80735	1.64908	1.8	1.64
0.9	1.9	1.8099	1.9	1.80999	1.90779	1.82022	1.9	1.81
1.0	2	1.9974	2	1.99984	2.00824	2.01135	2	2

$$y_1(x) = \frac{1}{8}x + \frac{17}{36} + \int_0^1 \frac{x+t}{3}(y_1(t) + y_2(t))dt,$$

$$y_2(x) = x^2 - \frac{19}{12}x + 1 + \int_0^1 xt(y_1(t) + y_2(t))dt,$$

with the exact solution $y_1(x) = x + 1$ *and* $y_2(x) = x^2 + 1$. *The comparison between the approximate solutions obtained by B-spline wavelet method for* $M = 2$ *and* $M = 4$ *and adaptive method based on trapezoidal rule [147] along with their exact solutions have been shown in Table 3.13. Table 3.14 cites the absolute errors obtained by these two methods.*

Example 3.4.3. *Consider the following system of Fredholm integral equations [126]*

TABLE 3.14: Absolute errors obtained by B-spline wavelet method and adaptive method based on trapezoidal rule for Example 3.4.2

x	B-spline wavelet method				Adaptive method	
	$M=2$		$M=4$			
	$y_1(x)$	$y_2(x)$	$y_1(x)$	$y_2(x)$	$y_1(x)$	$y_2(x)$
0.0	0	0.00260417	1.11022E-16	0.00016276	0.00378448	0
0.1	2.22045E-16	0.000104167	4.44089E-16	6.51042E-6	0.0042296	0.00113534
0.2	0	0.00114583	2.22045E-16	0.0000716146	0.00467471	0.00227069
0.3	2.22045E-16	0.00114583	2.22045E-16	0.0000716146	0.00511983	0.00340603
0.4	2.22045E-16	0.000104167	4.44089E-16	6.51042E-6	0.00556494	0.00454138
0.5	4.44089E-16	0.00260417	1.33227E-16	0.00016276	0.00601006	0.00567672
0.6	2.22045E-16	0.000104167	6.66134E-16	6.51042E-6	0.00645517	0.00681207
0.7	6.66134E-16	0.00114583	4.44089E-16	0.0000716146	0.00690029	0.00794741
0.8	4.44089E-16	0.00114583	4.44089E-16	0.0000716146	0.0073454	0.00908276
0.9	4.44089E-16	0.000104167	6.66134E-16	6.51042E-6	0.00779052	0.0102181
1.0	6.66134E-16	0.00260417	8.88178E-16	0.00016276	0.00823563	0.0113534

TABLE 3.15: Approximate solutions obtained by B-spline wavelet method and adaptive method based on trapezoidal rule along with exact solutions for Example 3.4.3

x	B-spline wavelet method				Adaptive method		Exact	
	$M=2$		$M=4$					
	$y_1(x)$	$y_2(x)$	$y_1(x)$	$y_2(x)$	$y_1(x)$	$y_2(x)$	$y_1(x)$	$y_2(x)$
0.0	-0.0026053	-0.002826	-0.000162	-0.000166	0.00067981	-0.000737	0	0
0.1	0.00989405	-0.0893165	0.00999348	-0.0890055	0.0108859	-0.0898863	0.01	-0.089
0.2	0.0411436	-0.150243	0.0400716	-0.151884	0.0410077	-0.152895	0.04	-0.152
0.3	0.0911431	-0.180735	0.0900716	-0.182865	0.0910608	-0.1838	0.09	-0.183
0.4	0.159893	-0.176037	0.159993	-0.176017	0.161061	-0.176636	0.16	-0.176
0.5	0.247392	-0.13151	0.249837	-0.125407	0.251023	-0.125427	0.25	-0.125
0.6	0.359892	-0.024483	0.359993	-0.0240153	0.360964	-0.0241933	0.36	-0.024
0.7	0.491142	0.136465	0.490072	0.133223	0.490898	0.133052	0.49	0.133
0.8	0.641141	0.355972	0.640072	0.352242	0.640842	0.352301	0.64	0.352
0.9	0.809891	0.638794	0.809993	0.638973	0.81081	0.639553	0.81	0.639
1.0	0.997391	0.989802	0.999837	0.999352	1.00082	1.00081	1	1

$$y_1(x) = f_1(x) + \int_0^1 (x-t)^3 y_1(t)dt + \int_0^1 (x-t)^2 y_2(t)dt,$$

$$y_2(x) = f_2(x) + \int_0^1 (x-t)^4 y_1(t)dt + \int_0^1 (x-t)^3 y_2(t)dt,$$

where $f_1(x) = \frac{3}{20} - \frac{11}{30}x - \frac{5}{3}x^2 + \frac{1}{3}x^3$ *and* $f_2(x) = -\frac{1}{30} + \frac{41}{60}x + \frac{3}{20}x^2 + \frac{23}{12}x^3 - \frac{1}{3}x^4$, *with the exact solution* $y_1(x) = x^2$ *and* $y_2(x) = x^3 + x^2 - x$. *The comparison between the approximate solutions obtained by B-spline wavelet method for* $M = 2$ *and* $M = 4$ *and adaptive method based on trapezoidal rule [147] along with their exact solutions have been shown in Table 3.15. Table 3.16 cites the absolute errors obtained by these two methods. From Table 3.16, it can be observed that the absolute errors obtained by the present method are very much better than that obtained by the adaptive method based on trapezoidal rule.*

TABLE 3.16: Absolute errors obtained by B-spline wavelet method and adaptive method based on trapezoidal rule for Example 3.4.3

x	B-spline wavelet method				Adaptive method	
	$M=2$		$M=4$			
	$y_1(x)$	$y_2(x)$	$y_1(x)$	$y_2(x)$	$y_1(x)$	$y_2(x)$
0.0	0.0.00260537	0.00282627	0.000162765	0.000166271	0.000679809	0.000737271
0.1	0.000105947	0.000316487	6.5174E-6	5.47084E-6	0.000885855	0.000886258
0.2	0.00114356	0.00175687	0.0000716057	0.000116058	0.00100765	0.000894502
0.3	0.00114313	0.00226496	0.000071604	0.000134611	0.00106078	0.000800344
0.4	0.000107237	0.0000370402	6.52242E-6	0.0000172473	0.00106081	0.000635888
0.5	0.00260757	0.00650994	0.000162774	0.000406899	0.00102334	0.000427009
0.6	0.000107887	0.000483014	6.52497E-6	0.0000153021	0.000963947	0.000193349
0.7	0.00114179	0.00346452	0.0000715988	0.000223463	0.000898205	0.0000516832
0.8	0.00114144	0.00397184	0.0000715974	0.000242013	0.000841698	0.000300911
0.9	0.000108955	0.000205924	6.52915E-6	0.0000270877	0.000810006	0.000553389
1.0	0.0026094	0.0101976	0.000162781	0.000647543	0.000818712	0.000814406

TABLE 3.17: Approximate solutions obtained by B-spline wavelet method and adaptive method based on trapezoidal rule for Example 3.4.4

x	B-spline wavelet method				Adaptive method	
	$M=2$		$M=4$			
	$y_1(x)$	$y_2(x)$	$y_1(x)$	$y_2(x)$	$y_1(x)$	$y_2(x)$
0.0	2.9777	1	2.98742	1	3.01611	1
0.1	2.59914	1.39881	2.59958	1.39881	2.62408	1.40161
0.2	2.29586	1.79761	2.29154	1.79761	2.31256	1.80322
0.3	2.06973	2.19642	2.06511	2.19642	2.08337	2.20483
0.4	1.92343	2.59522	1.9237	2.59523	1.93993	2.60644
0.5	1.86051	2.99403	1.87239	2.99403	1.88743	3.00805
0.6	1.91925	3.39283	1.92011	3.39284	1.93283	3.40966
0.7	2.08006	3.79164	2.07396	3.79165	2.08508	3.81127
0.8	2.35206	4.19045	2.3449	4.19045	2.35526	4.21288
0.9	2.74599	4.58925	2.74619	4.58926	2.75681	4.6145
1.0	3.27421	4.98806	3.29368	4.98807	3.30578	5.01611

Example 3.4.4. *Consider the following system of Fredholm integral equations*

$$y_1(x) = x + \int_0^1 |x-t|(y_1(t)+y_2(t))dt,$$

$$y_2(x) = 1 + x + \int_0^1 |xt|(y_1(t)+y_2(t))dt.$$

The exact solutions of this above system are not known. The approximate solutions obtained by B-spline wavelet method for $M = 2$ and $M = 4$ and adaptive method based on trapezoidal rule have been shown in Table 3.17. The Table 3.17 confirms that there is a good agreement of results between these two methods. Therefore, it verifies the ability, efficiency and applicability of the present method.

3.5 Application of B-spline wavelet method for solving system of nonlinear Fredholm integral equations of second kind

In this section, we consider the system of nonlinear Fredholm integral equations of second kind of the following form

$$\sum_{j=1}^{n} g_{i,j} y_j(x) = f_i(x) + \sum_{j=1}^{n} \int_0^1 K_{i,j}(x,t)\, F_{i,j}(t, y_j(t))\, dt, \quad i = 1, 2, ..., n, \tag{3.30}$$

where $f_i(x)$ and $K_{i,j}(x,t)$ are known functions and $y_j(x)$ are the unknown functions for $i, j = 1, 2, ...n$.

For solving eq. (3.30), first we assume

$$F_{i,j}(x, y_j(x)) = u_{i,j}(x), \quad 0 \le x \le 1. \tag{3.31}$$

Now from eq. (2.13) of Chapter 2, we can approximate the functions $u_{i,j}(x)$ and $y_j(x)$ as

$$u_{i,j}(x) = A_{i,j}^T \Psi(x), \tag{3.32}$$

$$y_j(x) = B_j^T \Psi(x), \tag{3.33}$$

where $A_{i,j}$ and B_j are $(2^{M+1} \times 1)$ column vectors similar to C defined in eq. (2.14) of Chapter 2.
Again by using dual wavelet functions, we can approximate the functions $f_j(x)$ and $K_{i,j}(x,t)$ as follows

$$f_i(x) = C_i^T \tilde{\Psi}(x), \tag{3.34}$$

$$K_{i,j}(x,t) = \tilde{\Psi}^T(t) \Theta_{i,j} \tilde{\Psi}(x), \tag{3.35}$$

where

$$\Theta_{i,j} = \int_0^1 \left[\int_0^1 K_{i,j}(x,t) \Psi(t) dt \right] \Psi^T(x) dx.$$

We can calculate C_i^T as

$$C_i^T = \int_0^1 f_i(x) \Psi^T(x) dx.$$

From the eqs. (3.31)-(3.35), we get

$$\begin{aligned}\int_0^1 K_{i,j}(x,t)F_{i,j}(t,y_j(t))dt &= \int_0^1 A_{i,j}^T\Psi(t)\tilde{\Psi}^T(t)\Theta_{i,j}\tilde{\Psi}(x)dt \\ &= A_{i,j}^T\left[\int_0^1 \Psi(t)\tilde{\Psi}^T(t)dt\right]\Theta_{i,j}\tilde{\Psi}(x) \\ &= A_{i,j}^T\Theta_{i,j}\tilde{\Psi}(x), \qquad (3.36) \\ \text{since} &\int_0^1 \Psi(t)\tilde{\Psi}^T(t)dt = I.\end{aligned}$$

Applying eq. (3.31)-(3.36) in eq. (3.30), we get

$$\sum_{j=1}^{n} g_{i,j}B_j^T\Psi(x) = C_i^T\tilde{\Psi}(x) + \sum_{j=1}^{n} A_{i,j}^T\Theta_{i,j}\tilde{\Psi}(x) \qquad (3.37)$$

Multiplying eq. (3.37) by $\Psi^T(x)$ both sides from the right and integrating with respect to x from 0 to 1, we have

$$\sum_{j=1}^{n} g_{i,j}B_j^T P = C_i^T + \sum_{j=1}^{n} A_{i,j}^T\Theta_{i,j}, \quad i = 1, 2, ..., n, \qquad (3.38)$$

where P is a $(2^{M+1}+1) \times (2^{M+1}+1)$ square matrix given by

$$P = \int_0^1 \Psi(x)\Psi^T(x)dx = \begin{bmatrix} P_1 & \\ & P_2 \end{bmatrix}$$

and

$$\int_0^1 \tilde{\Psi}(x)\Psi^T(x)dx = I.$$

Eq. (3.38) gives a system of $n(2^{M+1}+1)$ algebraic equations with $(n^2+n)(2^{M+1}+1)$ unknowns in $A_{i,j}$ and B_j for $i, j = 1, 2, ..., n$, given in eq. (3.32) and eq. (3.33).

To find the solutions $y_j(x)$ in eq. (3.33), we first utilize the following equations

$$F_{i,j}(x, B_j^T\Psi(x)) = A_{i,j}^T\Psi(x), \qquad (3.39)$$

with collocation points $x_s = \frac{s-1}{2^{M+1}}$, $s = 1, 2, ..., 2^{M+1}+1$.
Eq. (3.39) gives a system of $n^2(2^{M+1}+1)$ algebraic equations with $(n^2+n)(2^{M+1}+1)$ unknowns in $A_{i,j}$ and B_j for $i, j = 1, 2, ..., n$.

Combining eq. (3.38) and eq. (3.39), we have a total of $(n^2+n)(2^{M+1}+1)$ algebraic equations with $(n^2+n)(2^{M+1}+1)$ unknowns in $A_{i,j}$ and B_j for $i, j = 1, 2, ..., n$. Solving those equations for the unknown coefficients in the vectors $A_{i,j}$ and B_j for $i, j = 1, 2, ..., n$, we can obtain the solutions $y_j(x) = B_j^T\Psi(x)$, $j = 1, 2, ...n$.

TABLE 3.18: Approximate solutions obtained by B-spline wavelet method with exact solutions for Example 3.5.1

x	B-spline wavelet method				Exact	
	$M = 2$		$M = 4$			
	$y_1(x)$	$y_2(x)$	$y_1(x)$	$y_2(x)$	$y_1(x)$	$y_2(x)$
0	0	-0.00260699	0	-0.00016277	0	0
0.1	0.100542	0.00990657	0.100034	0.00999415	0.1	0.01
0.2	0.201084	0.0411905	0.200068	0.0400743	0.2	0.04
0.3	0.301625	0.0912447	0.300102	0.0900776	0.3	0.09
0.4	0.402167	0.160069	0.400136	0.160004	0.4	0.16
0.5	0.502709	0.247664	0.500169	0.249854	0.5	0.25
0.6	0.603251	0.360286	0.600203	0.360017	0.6	0.36
0.7	0.703792	0.491679	0.700237	0.490104	0.7	0.49
0.8	0.804334	0.641841	0.800271	0.640114	0.8	0.64
0.9	0.904876	0.810774	0.900305	0.810047	0.9	0.81
1	1.00542	0.998478	1.00034	0.999903	1	1

3.5.1 Illustrative examples

Example 3.5.1. *Consider the following system of Fredholm integral equations [208]*

$$y_1(x) = \frac{23}{35}x + \int_0^1 xt^2(y_1^2(t) + y_2^2(t))dt,$$

$$y_2(x) = \frac{11}{12}x + \int_0^1 x^2t(y_1^2(t) - y_2^2(t))dt,$$

with the exact solutions $y_1(x) = x$ and $y_2(x) = x^2$. The approximate solutions obtained by B-spline wavelet method for $M = 2$ and $M = 4$ with their exact solutions have been shown in Table 3.18 and Table 3.19 cites the absolute errors obtained by B-spline wavelet method.

Example 3.5.2. *Consider the following system of Fredholm integral equations [208]*

$$y_1(x) = 1 - \frac{17}{20}x - \frac{7}{6}x^2 + \int_0^1 xt^2y_1^3(t)dt + \int_0^1 x^2ty_2^2(t)dt,$$

$$y_2(x) = 1 - \frac{17}{12}x + x^2 - \frac{31}{10}x^3 + \int_0^1 xty_1^2(t)dt + \int_0^1 x^3ty_2^4(t)dt,$$

with the exact solutions $y_1(x) = x + 1$ and $y_2(x) = x^2 + 1$. The approximate solutions obtained by B-spline wavelet method for $M = 2$ and $M = 4$ with their exact solutions have been shown in Table 3.20 and Table 3.21 cites the absolute errors obtained by B-spline wavelet method.

TABLE 3.19: Absolute errors obtained by B-spline wavelet method for Example 3.5.1

x	B-spline wavelet method			
	$M=2$		$M=4$	
	$y_1(x)$	$y_2(x)$	$y_1(x)$	$y_2(x)$
0	0	0.00260699	0	0.000162771
0.1	0.000541752	0.0000934339	0.0000338869	5.84982E-6
0.2	0.0010835	0.00119046	0.0000677737	0.0000742656
0.3	0.00162526	0.00124469	0.000101661	0.0000775734
0.4	0.00216701	0.0000692524	0.000135547	4.07416E-6
0.5	0.00270876	0.00233585	0.000169434	0.000146232
0.6	0.00325051	0.000286167	0.000203321	0.0000173054
0.7	0.00379226	0.00167852	0.000237208	0.000104036
0.8	0.00433402	0.0018412	0.000271095	0.000113959
0.9	0.00487577	0.000774226	0.000304982	0.0000470757
1	0.00541752	0.00152242	0.000338869	0.0000966149

TABLE 3.20: Approximate solutions obtained by B-spline wavelet method with exact solutions for Example 3.5.2

x	B-spline wavelet method				Exact	
	$M=2$		$M=4$			
	$y_1(x)$	$y_2(x)$	$y_1(x)$	$y_2(x)$	$y_1(x)$	$y_2(x)$
0	1.00002	0.9974	1	0.999837	1	1
0.1	1.10059	1.01016	1.10004	1.01001	1.1	1.01
0.2	1.20104	1.04154	1.20007	1.0401	1.2	1.04
0.3	1.30135	1.09146	1.30009	1.0901	1.3	1.09
0.4	1.40153	1.15984	1.40011	1.15999	1.4	1.16
0.5	1.50159	1.24658	1.50011	1.24978	1.5	1.25
0.6	1.60147	1.35765	1.6001	1.35985	1.6	1.36
0.7	1.70144	1.48683	1.70009	1.48979	1.7	1.49
0.8	1.80083	1.63404	1.80006	1.63961	1.8	1.64
0.9	1.90032	1.7992	1.90003	1.80928	1.9	1.81
1	1.99968	1.98221	1.99999	1.99882	2	2

TABLE 3.21: Absolute errors obtained by B-spline wavelet method for Example 3.5.2

x	B-spline wavelet method			
	$M = 2$		$M = 4$	
	$y_1(x)$	$y_2(x)$	$y_1(x)$	$y_2(x)$
0	0.0000181015	0.00260009	0	0.00062756
0.1	0.000592624	0.000157298	0.0000405041	0.0000115058
0.2	0.00103682	0.0015415	0.000071711	0.000100139
0.3	0.00135068	0.00146444	0.000093696	0.0000962647
0.4	0.00153421	0.000159842	0.000106459	7.39085E-6
0.5	0.00158741	0.00341522	0.00011	0.000218003
0.6	0.00146683	0.00234858	0.000104139	0.000152962
0.7	0.00121593	0.0031679	0.0000890555	0.000209492
0.8	0.000834689	0.00595676	0.0000647503	0.000394768
0.9	0.000323121	0.0108014	0.0000312232	0.000715965
1	0.000318777	0.0177899	0.0000115258	0.00118026

Example 3.5.3. *Consider the following system of Fredholm integral equations [208]*

$$y_1(x) = -1 + \sec x - \tan 1 + \int_0^1 (y_1^2(t) + y_1(t)y_2(t))dt,$$

$$y_2(x) = 1 + \cos x - \tan 1 + \int_0^1 (y_1^2(t) - y_1(t)y_2(t))dt,$$

with the exact solutions $y_1(x) = \sec x$ and $y_2(x) = \cos x$. The approximate solutions obtained by B-spline wavelet method for $M = 2$ and $M = 4$ with their exact solutions have been shown in Table 3.22 and Table 3.23 cites the absolute errors obtained by B-spline wavelet method.

3.6 Error analysis

Theorem 3.6.1. *We assume that $f \in C^2[0,1]$ is represented by linear B-spline wavelets, where Ψ has two vanishing moments. Then $|d_{j,k}| \leq \alpha\beta\eta^2 \frac{2^{-3j}}{2!}$, where $\alpha = max|f''(t)|_{t\in[0,1]}$, $\beta = \int_{-k}^{2^j-k} \tilde{\psi}(x)dx$ and $\eta \in (-k, 2^j - k)$.*

Proof. Taylor expansion of $f \in C^2[0,1]$ about arbitrary $x_0 \in [0,1]$ can be written as

$$f(x) = f(x_0) + (x - x_0)f'(x_0) + \frac{(x - x_0)^2}{2!} f''(\xi), \quad \xi \in (0,1). \tag{3.40}$$

TABLE 3.22: Approximate solutions obtained by B-spline wavelet method with exact solutions for Example 3.5.3

x	B-spline wavelet method				Exact	
	$M = 2$		$M = 4$			
	$y_1(x)$	$y_2(x)$	$y_1(x)$	$y_2(x)$	$y_1(x)$	$y_2(x)$
0	0.99737	1.00239	0.999832	1.00015	1	1
0.1	1.00362	0.996143	1.00493	0.995077	1.00502	0.995004
0.2	1.01962	0.980591	1.02029	0.980101	1.02034	0.980067
0.3	1.04615	0.955883	1.04671	0.955372	1.04675	0.955336
0.4	1.08438	0.922212	1.08561	0.921133	1.0857	0.921061
0.5	1.03582	0.879817	1.13926	0.877724	1.13949	0.877583
0.6	1.20995	0.826451	1.21154	0.825336	1.21163	0.825336
0.7	1.3078	0.765485	1.30749	0.764842	1.30746	0.764842
0.8	1.4366	0.697407	1.43539	0.696707	1.43532	0.696707
0.9	1.6079	0.622757	1.6086	0.62161	1.60873	0.62161
1	1.83758	0.542128	1.84988	0.540302	1.85082	0.540302

TABLE 3.23: Absolute errors obtained by B-spline wavelet method for Example 3.5.3

x	B-spline wavelet method			
	$M = 2$		$M = 4$	
	$y_1(x)$	$y_2(x)$	$y_1(x)$	$y_2(x)$
0	0.0026298	0.0023933	0.000168178	0.000151052
0.1	0.00140108	0.00113913	0.0000899013	0.0000729561
0.2	0.000721751	0.000524883	0.0000470344	0.0000346196
0.3	0.000598287	0.00054679	0.0000426314	0.0000353846
0.4	0.00130308	0.00115059	0.0000929731	0.0000724765
0.5	0.0036784	0.00223398	0.000234847	0.000141089
0.6	0.00168167	0.00111533	0.000091301	0.0000726308
0.7	0.000343148	0.000642568	0.0000287158	0.0000424359
0.8	0.0012782	0.000699819	0.000069101	0.0000445436
0.9	0.000826292	0.00114689	0.000123569	0.0000713026
1	0.0132357	0.00182611	0.000934167	0.000114134

Now $f(x)$ can be presented by B-spline wavelets as

$$f(x) = C^T \Psi(x),$$

where

$$d_{j,k} = \int_0^1 f(x)\tilde{\psi}_{j,k}(x)dx. \tag{3.41}$$

Putting eq. (3.40) in eq. (3.41), we get

$$d_{j,k} = \int_0^1 f(x_0)\tilde{\psi}_{j,k}(x)dx + \int_0^1 (x - x_0)f'(x_0)\tilde{\psi}_{j,k}(x)dx + \int_0^1 \frac{(x - x_0)^2}{2!} f''(\xi)\tilde{\psi}_{j,k}(x)dx. \tag{3.42}$$

Putting $x_0 = \frac{k}{2^j}$ and $u = 2^j x - k$ in eq. (3.42), we have

$$d_{j,k} = 2^{-j} f(k/2^j) \int_{-k}^{2^j - k} \tilde{\psi}_{j,k}(u)du + 2^{-2j} f'(k/2^j) \int_{-k}^{2^j - k} u\tilde{\psi}_{j,k}(u)du + \frac{f''(\xi)}{2!} 2^{-3j} \int_{-k}^{2^j - k} u^2 \tilde{\psi}_{j,k}(u)du. \tag{3.43}$$

Suppose T is a linear transformation such that

$$T\psi = \tilde{\psi};$$

then taking the linear transformation T of first two integral of eq. (3.43) we have

$$d_{j,k} = 2^{-j} f(k/2^j) \int_{-k}^{2^j - k} T(\psi_{j,k}(u))du + 2^{-2j} f'(k/2^j) \int_{-k}^{2^j - k} uT(\psi_{j,k}(u))du + \frac{f''(\xi)}{2!} 2^{-3j} \int_{-k}^{2^j - k} u^2 \tilde{\psi}_{j,k}(u)du$$

$$d_{j,k} = 2^{-j} f(k/2^j) T\left(\int_{-k}^{2^j - k} \psi_{j,k}(u)du\right) + 2^{-2j} f'(k/2^j) T\left(\int_{-k}^{2^j - k} u\psi_{j,k}(u)du\right) + \frac{f''(\xi)}{2!} 2^{-3j} \int_{-k}^{2^j - k} u^2 \tilde{\psi}_{j,k}(u)du. \tag{3.44}$$

According to vanishing moments of order m, i.e.

$$\int_{-\infty}^{\infty} x^p \psi(x)dx = 0, \quad p = 0, 1, ..., m - 1,$$

the first two integrals of eq. (3.44) are zero. Then we have

$$d_{j,k} = \frac{f''(\xi)}{2!}2^{-3j}\int_{-k}^{2^j-k} u^2\tilde{\psi}_{j,k}(u)du. \tag{3.45}$$

Applying the mean value theorem for integral in eq. (3.45), we have

$$d_{j,k} = \frac{f''(\xi)}{2!}2^{-3j}\eta^2\int_{-k}^{2^j-k} \tilde{\psi}_{j,k}(u)du, \quad \eta \in (-k, 2^j - k).$$

Hence

$$|d_{j,k}| \leq \alpha\beta\eta^2\frac{2^{-3j}}{2!}.$$

□

Theorem 3.6.2. *Consider the Theorem 3.6.1 and assume that $e_j(x)$ is the error of approximation in V_j, then*

$$|e_j(x)| = O(2^{-2j}).$$

Proof. Any function $f(x) = L^2[0,1]$ can be approximated by linear B-spline wavelets as

$$f(x) = \sum_{k=-1}^{3} c_k\varphi_{2,k} + \sum_{i=2}^{\infty}\sum_{j=-1}^{2^i-2} d_{i,j}\psi_{i,j}. \tag{3.46}$$

If the above function truncated at M, then

$$f(x) \cong f^*(x) = \sum_{k=-1}^{3} c_k\varphi_{2,k} + \sum_{i=2}^{M}\sum_{j=-1}^{2^i-2} d_{i,j}\psi_{i,j}. \tag{3.47}$$

From eq. (3.46) and eq. (3.47), the error term can be calculated as (without loss of generality)

$$e_j(x) = \sum_{l=j}^{\infty}\sum_{k=-1}^{2^l-2} d_{l,k}\psi_{l,k}. \tag{3.48}$$

Setting

$$C_l = Max\left\{|\psi_{l,k}(x)|, \quad k = -1, ..., 2^l - 2\right\}. \tag{3.49}$$

Using Theorem 3.6.1 together with eq. (3.49), we obtain

$$|d_{l,k}\psi_{l,k}(x)| \leq \alpha\beta\eta^2 C_l\frac{2^{-3l}}{2!}.$$

This implies

$$\sum_{k=-1}^{2^l-2} |d_{l,k}\psi_{l,k}(x)| \leq \alpha\beta\eta^2 C_l\frac{2^{-2l}}{2!}.$$

Therefore, from eq. (3.48), we have

$$|e_j(x)| \leq \frac{\alpha\beta}{2!}\eta^2 \sum_{l=j}^{\infty} C_l 2^{-2l}.$$

Hence

$$|e_j(x)| = O(2^{-2j}).$$

□

3.7 Conclusion

In this chapter, semi-orthogonal compactly supported linear B-spline wavelets have been applied to find the numerical solution of linear and nonlinear Fredholm integral equations of second kind and their systems. The dual wavelets for these B-spline wavelets have been also presented. Because of semi-orthogonality, compact support and vanishing moments properties of B-spline wavelets, the matrices are very sparse. Using this procedure, the integral equations and their system have been reduced to solve systems of algebraic equations. In Section 3.2, linear Fredholm integral equations of second kind have been solved by using second order B-spline wavelets. The illustrative Examples 3.2.1-3.2.3 have been included to demonstrate the validity and applicability of the technique. These examples show the accuracy and efficiency of the described method. In Section 3.3, the approximate solution of the nonlinear Fredholm integral equation has been derived by using B-spline wavelet method and then compared with the VIM solution and exact solution. The obtained results are found to be in good agreement with the B-spline wavelet solutions. In Sections 3.4 and 3.5, the system of linear and nonlinear Fredholm integral equations has been solved by B-spline wavelet method. Additionally, in Section 3.4, the B-spline wavelet method has been compared with the adaptive method based on trapezoidal rule which confirms the plausibility of the B-spline technique. The obtained approximate solutions highly agree with the exact solutions. The illustrative examples have been included to demonstrate the validity and applicability of the technique. These examples also exhibit the accuracy and efficiency of the present method.

Chapter 4

Numerical Solutions of Nonlinear Fredholm Integral Equations System by Polynomial Approximation and Orthogonal Functions

4.1 Introduction

This chapter presents the approximation by polynomials and orthogonal functions for determining the solution of integral equations. Nonlinear integral equations appear in many problems of physical phenomena and engineering [208]. In recent years, many different polynomials and basic functions have been used to estimate the solution of integral equations. The approximate solutions for a system of nonlinear Fredholm integral equations of second kind are available in open literature. The learned researchers Biazer et al. have solved the system of nonlinear Fredholm integral equations of second kind by the Adomian decomposition method [22] and Legendre wavelets method [38]. In [19], the nonlinear Volterra integral equations system has been solved by Adomian decomposition method. Nonlinear Fredholm-Voltera integral equations system has been solved by homotopy perturbation method [223]. Hammerstein integral equations have been solved by Legendre approximation [199]. The system of nonlinear Fredholm-Hammerstein integral equations has been solved by B-spline wavelet method [180] in a previous chapter. In this chapter, we have implemented Bernstein polynomials and hybrid Legendre Block-Pulse functions to approximate the solution of the nonlinear Fredholm integral equations system. Since, the polynomials are differentiable and integrable, the Bernstein polynomials are defined on an interval to form a complete basis over the finite interval. On the other hand, each basis function of hybrid Legendre-Block-Pulse functions are piecewise continuous functions and also, these functions are orthonormal. It approximates any function defined on the interval $[0, 1]$ very accurately. The numerical technique based on hybrid Legendre-Block-Pulse function has been developed to approximate the solution of the system of nonlinear Fredholm-Hammerstein integral equations. These functions are formed by the hybridization of Legendre polynomials and Block-Pulse functions. These functions are orthonormal and have compact support on $[0, 1]$. The numerical results obtained by the present method have been compared with other methods. These proposed methods reduce the system of integral equations to a system of algebraic equations that can be solved easily by any of the usual numerical methods. Numerical examples are presented to illustrate the accuracy of the method.

4.2 Bernstein polynomial collocation method for solving nonlinear Fredholm integral equations system

Numerical methods based on the Bernstein polynomial have been developed to solve the integral equations of different types. The first time, the Bernstein polynomials have been used for the solution of some linear and nonlinear differential equations [33, 36, 220]. Mandal and Bhattacharya [138] obtained approximate solutions of some classes of integral equations by using Bernstein polynomials. Also, they used these polynomials to approximate the solution of linear Volterra integral equations [35] and a singular integro-differential equation [34]. The numerical solution of the system of linear Fredholm integral equations has been solved by using the Bernstein collocation method in [86]. In this section, we are going to propose a numerical approach to determine the solutions of a system of nonlinear Fredholm integral equations of second kind. This method is based on the Bernstein polynomial basis that approximates the unknown functions present in the integral equations. In this section, we consider the system of nonlinear Fredholm integral equations of second kind of the following form:

$$\sum_{j=1}^{n} g_{i,j}(x)y_j(x) = f_i(x) + \sum_{j=1}^{n}\int_0^1 K_{i,j}(x,t)F_{i,j}(t,y_j(t))dt, \quad i=1,2,...,n \tag{4.1}$$

where $f_i(x)$ and $K_{i,j}(x,t)$ are known functions and $y_j(x)$ are the unknown functions for $i,j = 1,2,...,n$. By applying the Bernstein collocation method (BCM), the integral equation defined in (4.1) reduces to a system of algebraic equations that can be solved easily by any numerical method. The numerical results are then compared with the results obtained by B-spline wavelet method (BWM). B-spline wavelets and Bernstein polynomials, and their properties and function approximations are defined in Chapter 2.

4.2.1 Solution to nonlinear Fredholm integral equations system by Bernstein collocation method

We have solved the system of nonlinear Fredholm integral equations of second kind of the form given in (4.1) by using Bernstein polynomials.

First, we assume

$$F_{i,j}(x,y_j(x)) = u_{i,j}(x), \qquad 0 \le x \le 1. \tag{4.2}$$

Now from eq. (2.68) of Chapter 2, we can approximate the functions $u_{i,j}(x)$ and $y_j(x)$ as

$$u_{i,j}(x) = A_{i,j}^T B(x), \tag{4.3}$$

$$y_j(x) = C_j^T B(x), \tag{4.4}$$

where $A_{i,j}$ and C_j are $(m+1)\times 1$ column vectors similar to C defined in Section 2.7.2 of Chapter 2.

Apply eqs. (4.2)-(4.4) in eq. (4.1), and we have

$$\sum_{j=1}^{n} g_{i,j}C_j^T B(x) = f_i(x) + \sum_{j=1}^{n}\int_0^1 K_{i,j}(x,t)A_{i,j}^T B(t)dt, \quad i=1,2,...,n.$$

$$\sum_{j=1}^{n} g_{i,j}C_j^T B(x) = f_i(x) + \sum_{j=1}^{n} A_{i,j}^T\int_0^1 K_{i,j}(x,t)B(t)dt, \quad i=1,2,...,n. \tag{4.5}$$

In eq. (4.5), the integral term $\int_0^1 K_{i,j}(x,t)B(t)dt$ is a function of x only. Let us assume this term as follows:

$$\int_0^1 K_{i,j}(x,t)B(t)dt = G_{i,j}(x). \tag{4.6}$$

Putting eq. (4.6) in eq. (4.5), we have

$$\sum_{j=1}^{n} g_{i,j}C_j^T B(x) = f_i(x) + \sum_{j=1}^{n} A_{i,j}^T G_{i,j}(x), \quad i = 1, 2, ..., n. \tag{4.7}$$

Using the collocation points $x_l = x_0 + lh$, where $x_0 = 0$, $h = \frac{1}{m}$ and $l = 0, 1, ..., m$, in eq. (4.7), we have

$$\sum_{j=1}^{n} g_{i,j}C_j^T B(x_l) = f_i(x_l) + \sum_{j=1}^{n} A_{i,j}^T G_{i,j}(x_l), \quad i = 1, 2, ..., n, \quad l = 0, 1, ..., m. \tag{4.8}$$

Now eq. (4.8) produces a system of algebraic equations with $n(m+1)$ equations along with $(n^2+n)(m+1)$ unknowns for the column vectors C_j and $A_{i,j}$.

Again consider the eq. (4.2) and utilizing it with the collocation points $x_l = x_0 + lh$, where $x_0 = 0$, $h = \frac{1}{m}$ and $l = 0, 1, ..., m$, we have

$$F_{i,j}(x_l, y_j(x_l)) = u_{i,j}(x_l), \quad i, j = 1, 2, ..., n; \quad l = 0, 1, ..., m. \tag{4.9}$$

Combining eqs. (4.8) and (4.9), we have a system of $(n^2 + n)(m + 1)$ algebraic equations along with the same number of unknowns. Solving this system by Newton's method, we can obtain the approximate values of column vectors C_j and $A_{i,j}$, $i, j = 1, 2, ..., n$. Hence we can obtain the approximate solutions of integral equations defined in eq. (4.1), i.e., $y_j(x) = C_j^T B(x)$, for $j = 1, 2, ..., n$.

4.2.2 Error analysis

Theorem 4.2.1. *Let $y_j(x)$, $j = 1, 2, ..., n$ be the solutions of the system of nonlinear Fredholm integral equations of second kind in eq. (4.1) and $Y_{j,m}(x) = \sum_{p=0}^{m} c_{j,p}B_{p,m}(x)$, $j = 1, 2, ..., n$ be the approximate solutions of the above integral equations system, where $B_{p.m}(x)$ is the Bernstein polynomials of degree m. Then the error term $\|e_m(x)\| \to 0$, if the above approximate polynomials converge to the exact solutions of the nonlinear Fredholm integral equations system, when $m \to \infty$.*

Proof. Consider $y_j(x)$, $j = 1, 2, ..., n$ to be the solutions of the integral equations system defined in eq. (4.1). Let $Y_{j,m}(x) = \sum_{p=0}^{m} c_{j,p}B_{p,m}(x)$,

$j = 1, 2, ..., n$ be the approximate solutions of eq. (4.1), i.e.,

$$\sum_{j=1}^{n} g_{i,j} Y_{j,m}(x) = f_i(x) + \sum_{j=1}^{n} \int_0^1 K_{i,j}(x,t) F_{i,j}(t, Y_{j,m}(t)) dt, \quad i = 1, 2, ..., n, \tag{4.10}$$

and it holds that

$$y_j(x) = \lim_{m \to \infty} Y_{j,m}(x). \tag{4.11}$$

Now define the error functions $e_m(x)$ by subtracting eq. (4.10) from eq. (4.1) as follows:

$$e_m(x) = \sum_{i=1}^{n} e_{i,m}(x), \tag{4.12}$$

where

$$e_{i,m}(x) = \sum_{i=1}^{n} g_{i,j}(y_j(x) - Y_{j,m}(x)) - \sum_{j=1}^{n} \int_0^1 K_{i,j}(x,t)(F_{i,j}(t, y_j(t)) - F_{i,j}(t, Y_{j,m}(t)))dt, \qquad i = 1, 2, ..., n.$$

From eq. (4.12), we have

$$\begin{aligned} \|e_m(x)\| &\leq \sum_{i=1}^{n} \|e_{i,m}(x)\| \\ &\leq \sum_{i=1}^{n} \sum_{j=1}^{n} \|g_{i,j}\| \|y_j(x) - Y_{j,m}(x)\| \\ &+ \sum_{i=1}^{n} \sum_{j=1}^{n} \int_0^1 \|K_{i,j}(x,t)\| \|F_{i,j}(t, y_j(t)) - F_{i,j}(t, Y_{j,m}(t))\| dt. \end{aligned}$$

Now, $\|g_{i,j}\|$, $\|K_{i,j}\|$ and $F_{i,j}$ are bounded.

And from eq. (4.11), $\|y_j(x) - Y_{j,m}(x)\| \to 0$ as $m \to +\infty$, and

$$\|F_{i,j}(t, y_j(t)) - F_{i,j}(t, Y_{j,m}(t))\| \leq L \|y_j(x) - Y_{j,m}(x)\| \to 0 \;\; as \;\; m \to +\infty,$$

where L is the Lipschitz constant. Hence, $\|e_m(x)\| \to 0$ as $m \to +\infty$ □

4.2.3 Illustrative examples

Example 4.2.1. *Consider the following system of Fredholm integral equations*

TABLE 4.1: Approximate solutions obtained by Bernstein collocation method and B-spline wavelet method along with their corresponding exact solutions for Example 4.2.1

x	Exact		BCM		BWM of order 4	
	$y_1(x)$	$y_2(x)$	$y_1(x)$	$y_2(x)$	$y_1(x)$	$y_2(x)$
0.0	0.0	0.00	0.0	0.00	0.000000	-0.00016277
0.1	0.1	0.01	0.1	0.01	0.100034	0.00999415
0.2	0.2	0.04	0.2	0.04	0.200068	0.0400743
0.3	0.3	0.09	0.3	0.09	0.300102	0.090076
0.4	0.4	0.16	0.4	0.16	0.400136	0.160004
0.5	0.5	0.25	0.5	0.25	0.500169	0.249854
0.6	0.6	0.36	0.6	0.36	0.600203	0.360017
0.7	0.7	0.49	0.7	0.49	0.700237	0.490104
0.8	0.8	0.64	0.8	0.64	0.800271	0.640114
0.9	0.9	0.81	0.9	0.81	0.900305	0.810047
1.0	1.0	1.00	1.0	1.00	1.00034	0.999903

$$\begin{aligned} y_1(x) &= \frac{23}{35}x + \int_0^1 xt^2(y_1^2(t) + y_2^2(t))dt, \\ y_2(x) &= \frac{11}{12}x^2 + \int_0^1 x^2t(y_1^2(t) - y_2^2(t))dt, \end{aligned}$$

with the exact solutions $y_1(x) = x$ and $y_2(x) = x^2$. The approximate solutions obtained by Bernstein polynomials of degree 10 *and B-spline wavelet method (BWM) of order* 4 *have been compared with their corresponding exact solutions in Table 4.1. Table 4.2 cites the absolute errors for the above approximate methods.*

Example 4.2.2. *Consider the following system of Fredholm integral equations*

$$\begin{aligned} y_1(x) &= 1 - \frac{17}{20}x - \frac{7}{6}x^2 + \int_0^1 xt^2y_1^3(t)dt + \int_0^1 x^2ty_2^2(t)dt, \\ y_2(x) &= 1 - \frac{17}{12}x + x^2 - \frac{31}{10}x^3 + \int_0^1 xty_1^2(t)dt + \int_0^1 x^3ty_2^4(t)dt, \end{aligned}$$

with the exact solutions $y_1(x) = x + 1$ and $y_2(x) = x^2 + 1$. The approximate solutions obtained by Bernstein polynomials of degree 10 *and B-spline wavelet method of order* 4 *have been compared with their corresponding exact solutions in Table 4.3. Table 4.4 cites the absolute errors for the above approximate methods.*

TABLE 4.2: Absolute errors with regard to Bernstein collocation method and B-spline wavelet method for Example 4.2.1

x	BCM		BWM of order 4	
	$y_1(x)$	$y_2(x)$	$y_1(x)$	$y_2(x)$
0.0	0	0	0	0.000162771
0.1	2.77556E-17	1.73472E-18	0.00003389	5.84982E-6
0.2	1.11022E-16	0	0.00006778	0.0000742656
0.3	5.55112E-17	4.16334E-17	0.000101661	0.0000775734
0.4	5.55112E-17	8.32667E-17	0.000135547	4.07416E-6
0.5	5.55112E-17	1.11022E-16	0.000169434	0.000146232
0.6	1.11022E-16	1.66533E-16	0.000203321	0.0000173054
0.7	1.11022E-16	2.22045E-16	0.000237208	0.000104036
0.8	1.11022E-16	2.22045E-16	0.000271095	0.000113959
0.9	0	4.44089E-16	0.000304982	0.0000470757
1.0	1.11022E-16	4.44089E-16	0.000338869	0.0000966149

TABLE 4.3: Approximate solutions obtained by Bernstein collocation method and B-spline wavelet method along with their corresponding exact solutions for Example 4.2.2

x	Exact		BCM		BWM of order 4	
	$y_1(x)$	$y_2(x)$	$y_1(x)$	$y_2(x)$	$y_1(x)$	$y_2(x)$
0.0	1.0	1.00	1.0	1.00	1	0.999837
0.1	1.1	1.01	1.1	1.01	1.10004	1.01001
0.2	1.2	1.04	1.2	1.04	1.20007	1.0401
0.3	1.3	1.09	1.3	1.09	1.30009	1.0901
0.4	1.4	1.16	1.4	1.16	1.40011	1.15999
0.5	1.5	1.25	1.5	1.25	1.50011	1.24978
0.6	1.6	1.36	1.6	1.36	1.6001	1.35985
0.7	1.7	1.49	1.7	1.49	1.70009	1.48979
0.8	1.8	1.64	1.8	1.64	1.80006	1.63961
0.9	1.9	1.81	1.9	1.81	1.90003	1.80928
1.0	2.0	2.00	2.0	2.00	1.99999	1.99882

TABLE 4.4: Absolute errors with regard to Bernstein collocation method and B-spline wavelet method for Example 4.2.2

x	BCM		BWM of order 4	
	$y_1(x)$	$y_2(x)$	$y_1(x)$	$y_2(x)$
0.0	0	0	0	0.00062756
0.1	0	4.44089E-16	0.0000405041	0.0000115058
0.2	6.66134E-16	6.66134E-16	0.000071711	0.000100139
0.3	2.22045E-16	2.22045E-16	0.000093696	0.0000962647
0.4	2.22045E-16	2.22045E-16	0.000106459	7.39085E-6
0.5	0	0	0.00011	0.000218003
0.6	0	2.22045E-16	0.000104139	0.000152962
0.7	2.22045E-16	2.22045E-16	0.0000890555	0.000209492
0.8	2.22045E-16	0	0.0000647503	0.000394768
0.9	0	0	0.0000312232	0.000715965
1.0	0	0	0.0000115258	0.00118026

Example 4.2.3. *Consider the following system of Fredholm integral equations*

$$y_1(x) = -1 + \sec x - \tan 1 + \int_0^1 (y_1^2(t) + y_1(t)y_2(t))dt,$$

$$y_2(x) = 1 + \cos x - \tan 1 + \int_0^1 (y_1^2(t) - y_1(t)y_2(t))dt,$$

with the exact solutions $y_1(x) = \sec x$ *and* $y_2(x) = \cos x$. *The approximate solutions obtained by Bernstein polynomials of degree* 10 *and B-spline wavelet method of order* 4 *have been compared with their corresponding exact solutions in Table 4.5. Table 4.6 cites the absolute errors for the above approximate methods.*

From these Tables, it is confirmed that the Bernstein collocation method gives a better approximation in comparison to the B-spline wavelet method.

4.3 Hybrid Legendre Block-Pulse functions for solving nonlinear Fredholm integral equations system

Hybrid Legendre-Block-Pulse functions are piecewise continuous functions. Also, these functions are orthonormal. It approximates any continuous function defined on the interval $[0, 1]$ very accurately. Numerical methods based on hybrid functions have been developed to solve the integral equations which

TABLE 4.5: Approximate solutions obtained by Bernstein collocation method and B-spline wavelet method along with their corresponding exact solutions for Example 4.2.3

x	Exact		BCM		BWM of order 4	
	$y_1(x)$	$y_2(x)$	$y_1(x)$	$y_2(x)$	$y_1(x)$	$y_2(x)$
0.0	1	1	0.999999	1	0.999832	1.00015
0.1	1.00502	0.995004	1.00502	0.995004	1.00493	0.995077
0.2	1.02034	0.980067	1.02034	0.980067	1.02029	0.980101
0.3	1.04675	0.955336	1.04675	0.955337	1.04671	0.955372
0.4	1.0857	0.921061	1.0857	0.921061	1.08561	0.921133
0.5	1.13949	0.877583	1.13949	0.877583	1.13926	0.877724
0.6	1.21163	0.825336	1.21163	0.825336	1.21154	0.825336
0.7	1.30746	0.764842	1.30746	0.764842	1.30749	0.764842
0.8	1.43532	0.696707	1.43532	0.696707	1.43539	0.696707
0.9	1.60873	0.62161	1.60872	0.62161	1.6086	0.62161
1.0	1.85082	0.540302	1.85081	0.540303	1.84988	0.540302

TABLE 4.6: Absolute errors with regard to Bernstein collocation method and B-spline wavelet method for Example 4.2.3

x	BCM		BWM of order 4	
	$y_1(x)$	$y_2(x)$	$y_1(x)$	$y_2(x)$
0.0	1.12723E-6	2.22986E-7	0.000168178	0.000151052
0.1	1.12723E-6	2.22986E-7	0.0000899013	0.0000729561
0.2	1.12723E-6	2.22986E-7	0.0000470344	0.0000346196
0.3	1.12723E-6	2.22986E-7	0.0000426314	0.0000353846
0.4	1.12723E-6	2.22986E-7	0.0000929731	0.0000724765
0.5	1.12723E-6	2.22986E-7	0.000234847	0.000141089
0.6	1.12723E-6	2.22986E-7	0.000091301	0.0000726308
0.7	1.12723E-6	2.22986E-7	0.0000287158	0.0000424359
0.8	1.12723E-6	2.22986E-7	0.000069101	0.0000445436
0.9	1.12723E-6	2.22986E-7	0.000123569	0.0000713024
1.0	1.12723E-6	2.22986E-7	0.000934167	0.000114134

are available in the literature. In [124], linear Fredholm integral equations of second kind have been solved by applying hybrid Legendre-Block-Pulse functions. Maleknejad et al., in [129], have solved Fredholm-Hammerstein integral equations by using this present method. Also many other hybrid functions have been used to solve integral equations. Hybrid Block-Pulse functions and the second Chebyshev polynomials [204] have been used to solve a system of integro-differential equations. In [131], hybrid orthonormal Bernstein and Block-Pulse functions have been applied to solve the Fredholm integral equations. Variational problems have been solved by using a hybrid of Block-Pulse functions and Bernoulli polynomials [161]. A hybrid of Fourier and Block-Pulse functions has been applied for solving integro-differential equations in [15]. In this section, we consider the system of nonlinear Fredholm integral equations of second kind the same as eq. (4.1). Hybrid Legendre Block-Pulse functions (HLBPF) have been used to solve the nonlinear Fredholm-Hammerstein integral equations system by reducing the integral equations system into a system of algebraic equations and that algebraic system has been solved numerically by Newton's method. The numerical results are then compared with the corresponding results obtained by the Legendre wavelet method (LWM).

4.3.1 Solution to nonlinear Fredholm integral equations system using hybrid Legendre Block-Pulse functions

In this section, we have solved the system of nonlinear Fredholm integral equations of second kind of the form given in eq. (4.1) by using hybrid Legendre Block-Pulse functions. First we assume

$$F_{i,j}(x, y_j(x)) = u_{i,j}(x), \quad 0 \le x \le 1. \tag{4.13}$$

Now from eq. (2.71) of Chapter 2, we can approximate the functions

$$u_{i,j}(x) = \Lambda_{i,j}^T B(x), \tag{4.14}$$

$$y_j(x) = X_j^T B(x), \tag{4.15}$$

$$f_i(x) = P_i^T B(x), \tag{4.16}$$

$$K_{i,j}(x) = B^T(t) T_{i,j} B(x), \tag{4.17}$$

where $\Lambda_{i,j}$, X_j and P_i are $(NM \times 1)$ column vectors defined similar to X defined in eq. (2.72) of Chapter 2. $B(x)$ is also $(NM \times 1)$ column vector as defined in eq. (2.73) of Chapter 2 and $T_{i,j}$ are $(NM \times NM)$ square matrices calculated as

$$T_{i,j} = \int_0^1 \left[\int_0^1 K_{i,j}(x,t) B(t) dt \right] B^T(x) dx, \quad i,j = 1, 2, ..., n,$$

and P_i can be calculated as

$$P_i = \int_0^1 f_i(x)B(x)dx, \quad i = 1, 2, ..., n.$$

Applying eqs. (4.13)-(4.17) in eq. (4.1), we have

$$\sum_{j=1}^{n} g_{i,j} X_j^T B(x) = P_i^T B(x) + \sum_{j=1}^{n} \int_0^1 \Lambda_{i,j}^T B(t) B^T(t) T_{i,j} B(x) dt, \quad i = 1, 2, ..., n.$$

$$\sum_{j=1}^{n} g_{i,j} X_j^T B(x) = P_i^T B(x) + \sum_{j=1}^{n} \Lambda_{i,j}^T T_{i,j} B(x), \quad i = 1, 2, ..., n. \tag{4.18}$$

Multiplying $B^T(x)$ on both sides of the eq. (4.18) from the right and integrating with respect to x from 0 to 1, we have

$$\sum_{j=0}^{n} g_{i,j} X_j^T = P_i^T + \sum_{j=1}^{n} \Lambda_{i,j}^T T_{i,j}, \quad i = 1, 2, ..., n, \tag{4.19}$$

since $\int_0^1 B(x)B^T(x)dx = I$.
Now eq. (4.19) gives a system of algebraic equations with NMn number of equations along with $NM(n^2 + n)$ unknowns for the column vectors X_j and $\Lambda_{i,j}$ defined in eq. (4.14) and eq. (4.15).

Again consider the equations defined in eq. (4.13) and utilizing it with the collocation points $x_s = \frac{2s-1}{2NM}$; $s = 1, 2, ..., NM$, we have

$$F_{i,j}(x_s, X_j^T B(x_s)) = \Lambda_{i,j}^T B(x_s), \quad i, j = 1, 2, ..., n; \quad s = 1, 2, ..., NM. \tag{4.20}$$

Now, eq. (4.20) gives a system of algebraic equations with NMn^2 equations along with $NM(n^2+n)$ unknowns for the column vectors X_j and $\Lambda_{i,j}$ defined in eq. (4.14) and eq. (4.15). Combining eqs. (4.19) and (4.20), we have a system of algebraic equations with $NM(n^2+n)$ equations along with the same number of unknowns. Solving this system, we can obtain the values of unknowns for the column vectors X_j and $\Lambda_{i,j}$, $i, j = 1, 2, ..., n$. Hence we can obtain the solutions of integral equations defined in eq. (4.1), i.e., $y_j(x) = X_j^T B(x)$, for $j = 1, 2, ..., n$.

4.3.2 Error analysis

Theorem 4.3.1. *The error term* $\|e(x)\|$ *obtained by hybrid Legendre Block-Pulse functions for solving a system of Fredholm-Hammerstein integral equations converges to zero as* $n \to \infty$ *for* $0 < \alpha < 1$, *where* $\alpha = |g_{i,j}^{-1}|KC_1$; $K = \sup_{0 \le x \le 1}\{\int_0^1 |K_{i,j}(x,t)|dt\}$ *and* C_1 *is Lipschitz constant.*

Proof. Let $y_{i,j}(l,m,x)$ be the approximate solutions of integral equations system defined in eq. (4.1) and $\|e_j(x)\|$ be the error term of each solution. Define

$$e(x) = \sum_j e_j(x). \tag{4.21}$$

Now,

$$\begin{aligned}\|e_j(x)\| &= \|y_j(x) - y_{l,m,x}\| \\ &\leq |g_{i,j}^{-1}| \int_0^1 \|K_{i,j}(x,t)\| \|F_{i,j}(t, y_j(t)) - F_{i,j}(t, y_j(l,m,t))\| dt. \end{aligned} \tag{4.22}$$

Suppose

$$K = \sup_{0 \leq x \leq 1} \{ \int_0^1 |K_{i,j}(x,t)| dt \}.$$

From eq. (4.22), we have

$$\begin{aligned}\|e_j(x)\| &\leq |g_{i,j}^{-1}| K \|F_{i,j}(t, y_j(t)) - F_{i,j}(t, y_j(l,m,t))\| \\ &= |g_{i,j}^{-1}| K C_1 \|y_j(t) - y_j(l,m,t)\|, \end{aligned} \tag{4.23}$$

where, $C_1 > 0$ is Lipschitz constant.

If we set

$$\alpha = |g_{i,j}^{-1}| K C_1,$$

then from eq. (4.23), we have

$$(1-\alpha)\|y_j(t) - y_j(l,m,t)\| \leq 0.$$

Now we choose α such that $0 < \alpha < 1$ and consequently

$$\|y_j(t) - y_j(l,m,t)\| \to 0 \quad \text{as} \quad n \to \infty,$$

which yields $\|e_j(x)\| \to 0$ as $n \to \infty$.
Hence from eq. (4.21), $\|e(x)\| \to 0$ as $n \to \infty$. □

4.3.3 Illustrative examples

Example 4.3.1. *Consider the following system of Fredholm integral equations*

$$\begin{aligned} y_1(x) &= \frac{23}{35}x + \int_0^1 xt^2(y_1^2(t) + y_2^2(t))dt, \\ y_2(x) &= \frac{11}{12}x^2 + \int_0^1 x^2 t(y_1^2(t) - y_2^2(t))dt, \end{aligned}$$

TABLE 4.7: Approximate solutions of $(y_1(x), y_2(x))$ obtained by HLBPF and LWM for Example 4.3.1

x	Exact	HLBPF		LWM	
		$M=3, N=4$	$M=4, N=8$	$M=3, k=3$	$M=4, k=4$
0.2	(0.2, 0.04)	(0.199987, 0.04)	(0.2, 0.04)	(0.199841, 0.0400073)	(0.2, 0.04)
0.4	(0.4, 0.16)	(0.399975, 0.16)	(0.4, 0.16)	(0.399682, 0.160029)	(0.4, 0.16)
0.6	(0.6, 0.36)	(0.599962, 0.36)	(0.6, 0.36)	(0.599522, 0.360066)	(0.599999, 0.36)
0.8	(0.8, 0.64)	(0.79995, 0.64)	(0.8, 0.64)	(0.799363, 0.640117)	(0.799999, 0.64)

TABLE 4.8: Absolute errors obtained by HLBPF for Example 4.3.1

x	Absolute errors of $(y_1(x), y_2(x))$ by HLBPF	
	$M=3, N=4$	$M=4, N=8$
0.2	(1.25015E-5, 1.18537E-8)	(3.80664E-8, 3.2843E-9)
0.4	(2.5003E-5, 4.74148E-8)	(7.61327E-8, 1.31372E-8)
0.6	(3.75045E-5, 1.06683E-7)	(1.14199E-7, 2.95587E-8)
0.8	(5.0006E-5, 1.89659E-7)	(1.52265E-7, 5.25489E-8)

with the exact solutions $y_1(x) = x$ and $y_2(x) = x^2$. The approximate solutions obtained by hybrid Legendre Block-Pulse functions for different values of N and M have been compared with the solutions obtained by Legendre wavelet method for different values of M and k and this comparison has been cited in Table 4.7. Table 4.8 cites the absolute errors for the above approximate solutions by HLBPF. The overall time estimation for Example 4.3.1 solved by HLBPF is 4.93 seconds and 30.076 seconds for $M = 3$, $N = 4$, and $M = 4$, $N = 8$, respectively. The memory space used for this computation is 37633540 bytes and 39593364 bytes for $M = 3$, $N = 4$, and $M = 4$, $N = 8$, respectively.

Example 4.3.2. *Consider the following system of Fredholm integral equations*

$$y_1(x) = 1 - \frac{17}{20}x - \frac{7}{6}x^2 + \int_0^1 xt^2 y_1^3(t)dt + \int_0^1 x^2 t y_2^2(t)dt,$$
$$y_2(x) = 1 - \frac{17}{12}x + x^2 - \frac{31}{10}x^3 + \int_0^1 xt y_1^2(t)dt + \int_0^1 x^3 t y_2^4(t)dt,$$

with the exact solutions $y_1(x) = x + 1$ and $y_2(x) = x^2 + 1$. The approximate solutions obtained by hybrid Legendre Block-Pulse functions for different values of N and M have been compared with the solutions obtained by Legendre wavelet method for different values of M and k and this comparison has been cited in Table 4.9. Table 4.10 cites the absolute errors for the above approximate solutions by HLBPF. The overall time estimation for Example 4.3.2 solved by HLBPF is 4.96 seconds and 29.094 seconds for $M = 3$, $N = 4$, and $M = 4$, $N = 8$, respectively. The memory space used for this computation is 37644356 bytes and 39543532 bytes for $M = 3$, $N = 4$, and $M = 4$, $N = 8$, respectively.

TABLE 4.9: Approximate solutions of $(y_1(x), y_2(x))$ obtained by HLBPF and LWM for Example 4.3.2

x	Exact	HLBPF		LWM	
		$M=3, N=4$	$M=4, N=8$	$M=3, k=3$	$M=4, k=4$
0.2	(1.2, 1.04)	(1.19996, 1.03998)	(1.2, 1.04)	(1.19934, 1.03968)	(1.2, 1.04)
0.4	(1.4, 1.16)	(1.39993, 1.15998)	(1.4, 1.16)	(1.3989, 1.15974)	(1.39999, 1.16)
0.6	(1.6, 1.36)	(1.59992, 1.36003)	(1.6, 1.36)	(1.5987, 1.3606)	(1.59999, 1.36)
0.8	(1.8, 1.64)	(1.79992, 1.64016)	(1.8, 1.64)	(1.79872, 1.64263)	(1.79999, 1.64001)

TABLE 4.10: Absolute errors obtained by HLBPF for Example 4.3.2

x	Absolute errors of $(y_1(x), y_2(x))$ by HLBPF	
	$M=3, N=4$	$M=4, N=8$
0.2	(4.10536E-5, 2.00077E-5)	(6.61903E-7, 3.44454E-7)
0.4	(6.78282E-5, 1.64768E-5)	(1.10299E-6, 3.30471E-7)
0.6	(8.03238E-5, 3.42853E-5)	(1.32326E-6, 4.0039E-7)
0.8	(7.85405E-5, 1.56357E-4)	(1.32272E-6, 2.20657E-6)

Example 4.3.3. *Consider the following system of Fredholm integral equations*

$$y_1(x) = -1 + \sec x - \tan 1 + \int_0^1 (y_1^2(t) + y_1(t)y_2(t))dt,$$
$$y_2(x) = 1 + \cos x - \tan 1 + \int_0^1 (y_1^2(t) - y_1(t)y_2(t))dt,$$

with the exact solutions $y_1(x) = \sec x$ and $y_2(x) = \cos x$. The approximate solutions obtained by hybrid Legendre Block-Pulse functions for different values of N and M have been compared with the solutions obtained by Legendre wavelet method for different values of M and k and this comparison has been cited in Table 4.11. Table 4.12 cites the absolute errors for the above approximate solutions by HLBPF. The overall time estimation for Example 4.3.3 solved by HLBPF is 71.277 seconds and 247.137 seconds for $M = 3$, $N = 4$, and $M = 4$, $N = 8$, respectively. The memory space used for this computation is 64347740 bytes and 70008212 bytes for $M = 3$, $N = 4$, and $M = 4$, $N = 8$, respectively.

Example 4.3.4. *Consider the following system of Fredholm integral equations*

$$y_1(x) = \mu \int_0^1 K(x,t) \left[(\beta - y_1(t))e^{y_1(t)} + (\beta - y_2(t))e^{y_2(t)}\right] dt,$$
$$y_2(x) = \int_0^1 \left(e^{xt}e^{y_1(t)} + e^{-xt}e^{y_2(t)}\right) dt,$$

TABLE 4.11: Approximate solutions of $(y_1(x), y_2(x))$ obtained by HLBPF and LWM for Example 4.3.3

x	Exact	HLBPF		LWM	
		$M=3, N=4$	$M=4, N=8$	$M=3, k=3$	$M=4, k=4$
0.2	(1.02034, 0.980067)	(1.02039, 0.980066)	(1.02034, 0.980066)	(1.02107, 0.979646)	(1.02035, 0.980063)
0.4	(1.0857, 0.921061)	(1.08581, 0.921066)	(1.08571, 0.92106)	(1.08649, 0.920646)	(1.08571, 0.921058)
0.6	(1.21163, 0.825336)	(1.21136, 0.825306)	(1.21163, 0.825335)	(1.21204, 0.824886)	(1.21164, 0.825333)
0.8	(1.43532, 0.696707)	(1.43425, 0.696664)	(1.43532, 0.696706)	(1.43492, 0.696244)	(1.43533, 0.696704)

TABLE 4.12: Absolute errors obtained by HLBPF for Example 4.3.3

x	Absolute errors of $(y_1(x), y_2(x))$ by HLBPF	
	$M=3, N=4$	$M=4, N=8$
0.2	(5.3713E-5, 5.89798E-7)	(1.46257E-6, 7.37282E-7)
0.4	(1.0432E-4, 5.47522E-6)	(2.47224E-6, 6.50458E-7)
0.6	(2.6703E-4, 2.9125E-5)	(3.09455E-6, 6.54156E-7)
0.8	(1.07747E-3, 4.27048E-5)	(1.39276E-6, 7.27628E-7)

where $K(x,t) = \begin{cases} e^{\lambda(x-t)}, & 0 \le x < t \\ 1, & t \le x \le 1 \end{cases}$, *and* $\mu = 0.02$, $\beta = 3$, *and* $\lambda = 10$. *The exact solutions of the above system for Fredholm integral equations are unknown. The approximate solutions obtained by hybrid Legendre Block-Pulse functions for different values of* N *and* M *have been compared with the solutions obtained by Legendre wavelet method for different values of* M *and* k *and cited in Table 4.13. The overall time estimation for Example 4.3.4 solved by HLBPF is 366.944 seconds and 2526.39 seconds for* $M = 3$, $N = 4$, *and* $M = 4$, $N = 8$, *respectively. The memory space used for this computation is 69862228 bytes and 129651640 bytes for* $M = 3$, $N = 4$, *and* $M = 4$, $N = 8$, *respectively. The Table 4.13 confirms that there is a good agreement of results between these two methods. Therefore it verifies the ability, efficiency and applicability of the present method.*

TABLE 4.13: Approximate solutions of $(y_1(x), y_2(x))$ obtained by HLBPF and LWM for Example 4.3.4

x	HLBPF		LWM	
	$M=3, N=4$	$M=4, N=8$	$M=3, k=3$	$M=4, k=4$
0.2	(0.0207894, 1.37402)	(0.0207893, 1.37401)	(0.0208479, 1.35188)	(0.0208482, 1.35187)
0.4	(0.0343434, 1.52401)	(0.0343436, 1.524)	(0.034531, 1.46505)	(0.0345316, 1.46504)
0.6	(0.0475951, 1.69596)	(0.0475884, 1.69597)	(0.0479556, 1.59915)	(0.0479491, 1.59917)
0.8	(0.0600656, 1.89354)	(0.0599892, 1.89357)	(0.0606086, 1.75719)	(0.0605317, 1.75722)

4.4 Conclusion

In this present analysis, we have proposed two computational methods to approximate the solutions of the system of nonlinear Fredholm integral equations of second kind. The proposed methods are very simple and straight forward, based on an approximation of the unknown function of an integral equation in terms of the Bernstein polynomials and hybrid Legendre Block-Pulse functions. Using this method, the system of integral equations has been reduced to solve a system of algebraic equations. In Section 4.2, the numerical solutions obtained by the Bernstein polynomial collocation method are compared with the B-spline wavelet method solutions. From the tables, it manifests that the present method has a better accuracy than the other method. For the first time, a nonlinear Fredholm-Hammerstein integral equations system has been solved by using hybrid Legendre Block-Pulse functions. The hybrid Legendre and Block-Pulse functions are orthogonal piecewise continuous functions which prompts flexibility for application. In Section 4.3, the numerical solutions obtained by HLBPF are compared with the LWM solutions. The overall cost estimation has been obtained for each problem and it verifies that the above said HLBPF method is not expensive and it can be applied for solving other integral and differential equations. The illustrated examples analyze and verifies the ability and the reliability of the present methods. Solutions obtained by the present methods compared to exact solutions admit a remarkable efficiency and accuracy. The main advantages of these methods are their efficiency and simple applicability.

Chapter 5

Numerical Solutions of Hammerstein Integral Equations Arising in Chemical Phenomena

5.1 Introduction

In this chapter, the numerical solutions of nonlinear Hammerstein integral equations (HIEs) arising in chemical phenomena are discussed. First, we have solved an integral equation which forms the basis for the conductor like screening model for real solvent (COSMO-RS) which appeared in chemical phenomena. The conductor like screening model for real solvent (COSMO-RS) [100] is a quantum chemistry based equilibrium thermodynamics method with the purpose of predicting chemical potential μ in liquids. It processes the screening charge density σ on the surface of molecules to calculate the chemical potential μ of each species in solution. As an initial step a quantum chemical COSMO calculation for all molecules is performed and the results (e.g. screening change density) are stored in a database. In a separate step COSMO-RS uses the stored COSMO results to calculate the chemical potential of the molecules in a liquid solvent or mixture. The resulting chemical potentials are the basis for other thermodynamic equilibrium properties such as activity co-efficients, solubility, partition co-efficients, vapor pressure and free energy of solvation. The method was developed to provide a general prediction method with no need for system specific adjustment. Due to use of σ from the COSMO calculation, COSMO-RS does not require functional group parameters. Quantum chemical effects like group-group interaction, mesmeric effects and inductive effects also are incorporated into COSMO-RS by this approach.

Our aim is to solve the Hammerstein nonlinear integral equation

$$\mu_S(\sigma) = -RT \ln \left[\int P_S(\sigma') \exp(-\frac{E_{int}(\sigma, \sigma') - \mu_S(\sigma')}{RT}) d\sigma' \right],$$

where R is the gas constant, T is the temperature and the term $E_{int}(\sigma, \sigma')$ denotes the interaction energy expression for the segments with screening charge density σ and σ' respectively, the molecular interaction in the solvent is $P_S(\sigma)$ and the chemical potential of the surface segments is described by $\mu_S(\sigma)$ which is to be determined. This problem has been solved by the Bernstein collocation method, the Haar wavelet method, and the Sinc collocation method. All the above methods have been applied to solve the integral equations by reducing to a system of algebraic equations. Comparison has been done for these methods, to demonstrate the validity and applicability of the Bernstein collocation method, the Haar wavelet method and the Sinc collocation method.

Second, the mathematical model has been considered for an adiabatic tubular chemical reactor which processes an irreversible exothermic chemical reaction. For a steady state solution for an adiabatic tubular chemical reactor, the model can be reduced to an ordinary differential equation with a parameter in the boundary conditions. For easy computation, the problem can be converted into a Hammerstein integral equation which can be solved easily. The numerical method based on linear B-spline wavelets has been developed to approximate the solution of the Hammerstein integral equation. This method reduces the integral equation to a system of algebraic equations that can be solved numerically easily. The numerical results obtained by the present method have been compared with the results obtained by Contraction mapping principle (CMP), Shooting method (SM) and Adomian's decomposition method (ADM) to demonstrate the validity and applicability of the present method.

5.2 Comparative experiment on the numerical solutions of Hammerstein integral equation arising from chemical phenomena

The COSMO-RS method was first published in 1995 by A. Klamt [100]. A refined version of COSMO-RS was published in 1998 [101] and is the basis for new developments and reimplementation [70, 28, 63, 127]. Within the basic formulation of COSMO-RS, interaction terms depend on the screening charge density σ. Each molecule and mixture can be represented by the histogram $P(\sigma)$, the so called σ-profile. The σ-profile of a mixture is the weighted sum of the profiles of all its components. Using the interaction energy $E_{int}(\sigma, \sigma')$ and the σ-profile of the solvent $P(\sigma')$, the chemical potential $\mu_S(\sigma)$ of a surface piece with screening charge σ is determined as [127]

$$\mu_S(\sigma) = -RT \ln \left[\int P_S(\sigma') \exp(-\frac{E_{int}(\sigma, \sigma') - \mu_S(\sigma')}{RT}) d\sigma' \right], \tag{5.1}$$

where R is the gas constant, T is the temperature and the term $E_{int}(\sigma, \sigma')$ denotes the interaction energy expression for the segments with screening charge density σ and σ' respectively, the molecular interaction in solvent is $P_S(\sigma)$ and the chemical potential of the surface segments is described by $\mu_S(\sigma)$ which is to be determined. The domain of integration is determined by the characteristics of the σ-profile.

We can rewrite the eq. (5.1) as

$$-\frac{\mu_S(\sigma)}{RT} = \ln \left[\int_a^b K(\sigma, \sigma') \exp\left(\frac{\mu_S(\sigma')}{RT}\right) d\sigma' \right],$$

where $K(\sigma, \sigma') = P_S(\sigma')\Omega(\sigma, \sigma')$ and $\Omega(\sigma, \sigma') = \exp\{-\frac{E_{int}(\sigma,\sigma')}{RT}\}$.
Now, by substituting $y(\sigma) = \exp\left(-\frac{\mu_S(\sigma)}{RT}\right)$, we have

$$y(\sigma) = \int_a^b K(\sigma, \sigma') \left(y(\sigma')\right)^{-1} d\sigma'. \tag{5.2}$$

The above eq. (5.2) is nothing but the well known nonlinear Hammerstein integral equation.
The general form of nonlinear Hammerstein integral equation is given as [179]

$$y(x) = g(x) + \int_a^b K(x, t)F(t, y(t))dt, \tag{5.3}$$

where $K(x, t)$, $g(x)$ and $F(t, y)$ are known functions and $y(x)$ is the unknown function which should be determined. A computational approach to solve an integral equation is an essential work in scientific research. There are many numerical methods available for solving Hammerstein integral equations [179, 176, 123, 107, 120, 160, 226]. The learned researchers Dehghan et al. have applied Bernstein polynomial operational matrices for solving age-structured population models [221]. The Ritz-Galerkin method with Bernstein polynomial basis [219] has been used for solving a heat equation with non-classic boundary conditions. The authors have applied the Bernstein polynomial to solve a system of nonlinear Fredholm integral equations [183]. The Thomas-Fermi equation [152] has been solved by the Sinc collocation method that converges to the solution at an exponential rate. Saadatmandi et al. [167] have solved a class of fractional convection-diffusion equations with variable coefficients by the Sinc-Legendre collocation method. The Haar wavelet method has been applied to solve fractional differential equations by Saha Ray et al. [172, 175]. Legendre multi-wavelets have been used for solving weakly singular Fredholm integro-differential equations [109]. Some iterative techniques and quadrature formulae [179, 176] have been applied to solve Hammerstein integral equations by Saha Ray et al. Also the Hammerstein integral equation has been solved by B-spline wavelets [107].

In this present study, nonlinear Hammerstein integral equations have been solved by the Bernstein collocation method, the Haar wavelet method, and the Sinc collocation method. All the above methods have been applied to solve the integral equations by reducing to a system of algebraic equations. From the obtained results, it is quite plausible that the results obtained by the Bernstein collocation method converge more rapidly than the other two methods.

5.2.1 Bernstein collocation method

Bernstein polynomials defined in Section 2.7 of Chapter 2 form a complete basis [183] over the interval $[a, b]$. It is easy to show that any given polynomials of degree n can be expressed in terms of a linear combination of the basis functions. A function $y(x)$ defined over $[a, b]$ can be approximated by Bernstein polynomials basis functions of degree n as

$$y(x) \approx \sum_{i=0}^{n} c_i B_{i,n}(x) = C^T B(x), \tag{5.4}$$

where C and $B(x)$ are $(n+1) \times 1$ column vectors defined in Section 2.7.2 of Chapter 2. For solving eq. (5.3), we first assume

$$z(x) = F(x, y(x)) = (y(x))^{-1}. \tag{5.5}$$

From eq. (5.3), we can approximate the unknown function $y(x)$ and $z(x)$ as

$$y(x) = C_1^T B(x), \tag{5.6}$$

$$z(x) = C_2^T B(x), \tag{5.7}$$

where C_1 and C_2 are $(n+1) \times 1$ column vectors similar to C. Substituting eqs. (5.5)-(5.7) in eq. (5.3), we have

$$C_1^T B(x) = g(x) + \int_a^b K(x,t) C_2^T B(t) dt. \tag{5.8}$$

Taking $G(x) = \int_a^b K(x,t)B(t)dt$ then the eq. (5.8) can be reduced as

$$C_1^T B(x) = g(x) + C_2^T G(x). \tag{5.9}$$

Again from eq. (5.5), we have

$$C_2^T B(x) = F(x, C_1^T B(x)). \tag{5.10}$$

From eqs. (5.9) and (5.10), we get

$$C_2^T B(x) = F(x, g(x) + C_2^T G(x)). \tag{5.11}$$

Putting the collocation points $x_l = x_0 + lh$, where $x_0 = a$, $h = \frac{b-a}{n}$ and $l = 0, 1, ..., n$, in eq. (5.11), we have

$$C_2^T B(x_l) = F(x_l, g(x_l) + C_2^T G(x_l)). \tag{5.12}$$

Now eq. (5.12) gives a system of $n+1$ algebraic equations with the same number of unknowns for vector C_2. Numerically solving this algebraic system, we can obtain the values of unknowns for vector C_2 and hence the solution $y(x) = f(C_2^T B(x)) = \frac{1}{C_2^T B(x)}$.

5.2.2 Haar wavelet method

The Haar functions are the family of switched rectangular waveforms where amplitudes can differ from one function to another function. Usually the Haar wavelets are defined on the interval $[0,1]$ but in a general case these are defined on $[a,b]$. We divide the interval $[a,b]$ into m equal subintervals. In this case the orthogonal set of Haar functions is defined on the interval $[a,b]$ by [172, 175] (see Section 2.5 of Chapter 2).

An arbitrary function $y(x) \in L^2[a,b]$ can be approximated with finite terms as

$$y(x) \approx \sum_{i=0}^{m-1} c_i h_i(x) = C_{(m)}^T h_{(m)}(x), \tag{5.13}$$

where the coefficient vector $C_{(m)}^T$ and the Haar function vector $h_{(m)}(x)$ are, respectively, defined in eq. (2.58) of Chapter 2.

For solving eq. (5.3), we first assume

$$z(x) = F(x, y(x)) = (y(x))^{-1}. \tag{5.14}$$

From eq. (5.3), we can approximate the unknown function $y(x)$ and $z(x)$as

$$y(x) = A^T_{(m)} h_{(m)}(x), \tag{5.15}$$

$$z(x) = B^T_{(m)} h_{(m)}(x), \tag{5.16}$$

where $A^T_{(m)}$ and $B^T_{(m)}$ are $1 \times (m+1)$ row vectors similar to $C^T_{(m)}$.
Substituting eqs. (5.14)-(5.16) in eq. (5.3), we have

$$A^T_{(m)} h_{(m)}(x) = g(x) + \int_a^b K(x,t) B^T_{(m)} h_{(m)}(t) dt. \tag{5.17}$$

Taking $G_{(m)}(x) = \int_a^b K(x,t) h_{(m)}(t) dt$ then the eq. (5.17) can be reduced as

$$A^T_{(m)} h_{(m)}(x) = g(x) + B^T_{(m)} G_{(m)}(x). \tag{5.18}$$

Again from eq. (5.14), we have

$$B^T_{(m)} h_{(m)}(x) = F(x, A^T_{(m)} h_{(m)}(x)). \tag{5.19}$$

From eqs. (5.18) and (5.19), we get

$$B^T_{(m)} h_{(m)}(x) = F(x, g(x) + B^T_{(m)} G_{(m)}(x)). \tag{5.20}$$

Putting the collocation points $x_l = a + (l - 0.5)\frac{(b-a)}{m}, \quad l = 1, 2, ..., m$ in eq. (5.20), we have

$$B^T_{(m)} h_{(m)}(x_l) = F(x_l, g(x_l) + B^T_{(m)} G_{(m)}(x_l)). \tag{5.21}$$

Now eq. (5.21) gives a system of m algebraic equations with the same number of unknowns for vector $B^T_{(m)}$. Numerically solving this algebraic system, we can obtain the values of unknowns for vector $B^T_{(m)}$ and hence the solution $y(x) = f(B^T_{(m)} h_{(m)}(x)) = \frac{1}{B^T_{(m)} h_{(m)}(x)}$.

5.2.3 Sinc collocation method

In this section, we have employed Sinc function approximation to solve the nonlinear Hammerstein integral equation. The properties of Sinc function and the function approximation are defined in Section 2.9 of Chapter 2.

Consider the nonlinear Hammerstein integral equation defined in eq. (5.3) as

$$y(x) = g(x) + \int_a^b K(x,t) F(t, y(t)) dt, \quad x \in [a, b]. \tag{5.22}$$

By applying the formula defined in eq. (2.76) of Chapter 2 in eq. (5.22), we have

$$\int_a^b K(x,t)F(t,y(t))dt \approx h \sum_{j=-N}^{N} K(x,t_j)F(t_j,y(t_j))\varphi'(jh), \tag{5.23}$$

where

$$h = \frac{1}{N}\ln\left(\frac{2\pi dN}{\alpha}\right),$$
$$t_j = \varphi(jh), \quad j = -N, ..., N.$$

From eqs. (5.22) and (5.23), we can reduce the integral equation to the algebraic system by putting $x = x_k$ as

$$y(x_k) = g(x_k) + h \sum_{j=-N}^{N} K(x_k,t_j)F(t_j,y(t_j))\varphi'(jh), \tag{5.24}$$

where $x_k = \varphi(kh)$, $k = -N, ..., N$.
Eq. (5.24) gives a system of $2N+1$ nonlinear algebraic equations along with unknowns $y_j = y(x_j)$, $\quad j = -N, ..., N$.
Solving this nonlinear system by Newton's method, we obtain the approximate solutions y_j, $\quad j = -N, ..., N$. Then we can obtain the approximate solution for eq. (5.3) as

$$y_N(x) = g(x) + h \sum_{j=-N}^{N} K(x,t_j)F(t_j,y_j)\varphi'(jh). \tag{5.25}$$

5.2.4 Illustrative examples

Example 5.2.1. *Let us now investigate the solution of the COSMO-RS integral equation [127]*

$$y(\sigma) = \int_a^b P_S(\sigma')\Omega(\sigma,\sigma')(y(\sigma'))^{-1}d\sigma'$$

for a particular case of the energy expression, namely the electrostatic misfit energy. In this case the relevant part of the kernel of the integral equation is given by $\Omega(\sigma,\sigma') = \exp\{-(\sigma+\sigma')^2\}$.
Also, we employ the following piecewise defined analytical function as synthetic σ*-profile:*

$$P_S(\sigma) = \begin{cases} \exp(-(5\sigma+2.5)^2) + \frac{1}{25\sigma^2+1} + \frac{(\sin(5\sigma+2.5))^2}{(5\sigma-2.5)^4+1} + q(5\sigma), & -2 \le \sigma < 2 \\ 0, & \textit{Otherwise} \end{cases}$$

where $q(\sigma) = \begin{cases} -(\sigma-7)(\sigma-9), & 7 \le \sigma < 9 \\ 0, & otherwise. \end{cases}$
So, COSMO-RS integral equation will be

$$y(\sigma) = \int_{-3}^{3} P_S(\sigma')\Omega(\sigma,\sigma')(y(\sigma'))^{-1} d\sigma'.$$

This problem has been solved by the above three numerical methods, viz. the Bernstein collocation method (BCM), the Haar wavelet method (HWM), and the Sinc collocation method (SCM). The numerical results obtained by the above three methods have been cited in Table 5.1. In this problem, the reduced algebraic system has been solved by Newton's method with initial guess at zero so that the Jacobian is nonsingular.

TABLE 5.1: Numerical results for Example 5.2.1

x	BCM			HWM			SCM		
	$n=20$	$n=40$	$n=60$	$m=32$	$m=64$	$m=128$	$N=20$	$N=40$	$N=60$
-3	0.33983	0.3399	0.33989	0.44725	0.38618	0.3624	0.41505	0.31313	0.34715
-2.5	1.0436	1.03996	1.037	1.00784	1.0684	1.02392	1.25782	0.95974	1.0586
-2	1.9687	1.96368	1.96735	2.06715	1.94737	1.9819	2.36792	1.82947	2.00682
-1.5	2.39346	2.39354	2.39345	2.42834	2.39556	2.39436	2.84492	2.24043	2.43812
-1	2.0452	2.0453	2.04523	2.11768	2.0326	2.05572	2.36157	1.93929	2.07666
-0.5	1.49406	1.49409	1.49405	1.47893	1.51148	1.48742	1.62531	1.45005	1.50733
0	1.10484	1.10484	1.10484	1.04319	1.07366	1.08914	1.12284	1.09871	1.10684
0.5	0.76102	0.76102	0.76102	0.78163	0.74959	0.76651	0.74178	0.76831	0.75897
1	0.41419	0.41419	0.41419	0.39251	0.42374	0.40917	0.39492	0.42196	0.41195
1.5	0.16382	0.16381	0.16382	0.13146	0.14740	0.15548	0.15365	0.16812	0.16255
2	0.04623	0.04623	0.04623	0.05002	0.04415	0.04725	0.04255	0.04789	0.04572
2.5	0.00967	0.00967	0.00967	0.00860	0.01018	0.00942	0.00865	0.01019	0.00950
3	0.00163	0.00163	0.00163	0.03384	0.03375	0.03367	0.00140	0.00177	0.00158

Example 5.2.2. *Consider the nonlinear Hammerstein integral equation* (5.3) *(see ref. [127]) where* $K(x,t) = e^{-10(x+t)}$, $F(t, y(t)) = (y(t))^{-1}$, *and* $g(x) = \frac{21-11\exp(10)}{100}\exp(-10(1+x)) + \frac{1}{1+x}$. *The interval of integration is* $[0,1]$. *The exact equation of this problem is* $y(x) = \frac{1}{1+x}$. *This problem has been solved by the above three numerical methods, viz. the Bernstein collocation method (BCM), the Haar wavelet method (HWM), and the Sinc collocation method (SCM) and the corresponding absolute errors have been cited in Table 5.2.*

5.3 Numerical solution for Hammerstein integral equation arising from chemical reactor theory by using semiorthogonal B-spline wavelets

In this section, a mathematical model has been developed for an adiabatic tubular chemical reactor [158] which processes an irreversible exothermic

TABLE 5.2: Absolute errors for Example 5.2.2

x	BCM ($n = 10$)	HWM ($m = 32$)	SCM ($N = 10$)
0	0	0.0153187	0.00775417
0.1	2.22045E-16	0.00765666	0.0028526
0.2	3.33067E-16	0.0021544	0.00104941
0.3	3.33067E-16	0.00185752	0.000386057
0.4	1.11022E-16	0.00481696	0.000142023
0.5	0	0.00687241	0.0000522472
0.6	1.11022E-16	0.0036406	0.0000192207
0.7	1.11022E-16	0.00107926	7.07089E-6
0.8	1.11022E-16	0.00096621	2.60123E-6
0.9	0	0.00260984	9.56941E-7
1	0	0.00393701	3.52039E-7

chemical reaction. For a steady state solution, the model can be reduced to an ordinary differential equation with a parameter in the boundary conditions [119] as follows

$$u'' - \lambda u' + F(\lambda, \mu, \beta, u) = 0, \tag{5.26}$$

with boundary conditions

$$u'(0) = \lambda u(0), \quad u'(1) = 0, \tag{5.27}$$

where

$$F(\lambda, \mu, \beta, u) = \lambda\mu(\beta - u)exp(u).$$

The unknown u represents the steady state temperature of the reaction, and the parameters λ, μ and β represent the Peclet number, the Damkohler number and the dimensionless adiabatic temperature rise respectively. This problem has been studied by many authors [158, 119, 78, 79]. The existence of a numerical solution of this problem for a particular parameter range has been discussed in [158, 119, 78, 79].

In order to develop the solution of the problem defined in eq. (5.26), the problem can be converted into a nonlinear Hammerstein integral equation by using Green's function. The Hammerstein integral form of eq. (5.26) with boundary conditions eq. (5.27) can be defined as

$$u(x) = \int_0^1 K(x,t)g(t,u(t))dt \qquad 0 \leq x \leq 1, \tag{5.28}$$

where

$$K(x,t) = \begin{cases} e^{\lambda(x-t)}, & \text{if } 0 \leq x \leq t, \\ 1, & \text{if } t \leq x \leq 1, \end{cases}$$

and

$$g(t, u(t)) = \mu(\beta - u)exp(u).$$

In this section, we consider eq. (5.28) as a Hammerstein integral equation in the space of continuous functions on the closed interval. Throughout, we assume λ and μ are positive, and β is nonnegative. Our main work is to solve this Hammerstein integral equation by B-spline wavelet method. B-spline wavelet method has been applied to solve the integral equations of different forms [179, 181, 180]. The learned researchers Saha Ray et al. have solved nonlinear Fredholm integral equations [179] and the system of linear and nonlinear Fredholm integral equations [181, 180] by B-spline wavelet method. The B-spline wavelet method converts the Hammerstein integral equation to a system of algebraic equations and that algebraic equations system again can be solved by any of the usual numerical methods. The obtained results have been compared with the results obtained by Adomian's decomposition method, contraction mapping principle and shooting method [119].

5.3.1 Application of B-spline wavelet method to the Hammerstein integral equations

In this section, we have solved the nonlinear Fredholm Hammerstein integral equation defined in eq. (5.28) using B-spline wavelets. B-spline wavelets and their function approximations are defined in Section 2.2 of Chapter 2. First, we assume

$$g(x, u(x)) = z(x) \quad 0 \leq x \leq 1. \tag{5.29}$$

Now from eq. (2.13) of Chapter 2, we can approximate the functions $z(x)$ and $u(x)$ as

$$z(x) = A^T \Psi(x), \tag{5.30}$$

$$u(x) = B^T \Psi(x), \tag{5.31}$$

where A and B are $(2^{M+1}+1) \times 1$ column vectors similar to C as in eq. (2.14) of Chapter 2.
Again using the dual wavelet functions, we can approximate $K(x,t)$ as follows.

$$K(x,t) = \tilde{\Psi}^T(x)\Theta\tilde{\Psi}(x), \tag{5.32}$$

where

$$\Theta = \int_0^1 \left[\int_0^1 K(x,t)\Psi(t)dt\right] \Psi(x)dx.$$

From eqs. (5.29)-(5.32), we get

$$\begin{aligned}\int_0^1 K(x,t)g(t,u(t))dt &= \int_0^1 A^T\Psi(t)\tilde{\Psi}^T(t)\Theta\tilde{\Psi}(x)dt \\ &= A^T\left[\int_0^1 \Psi(t)\tilde{\Psi}^T(t)dt\right]\Theta\tilde{\Psi}(x) \\ &= A^T\Theta\tilde{\Psi}(x),\end{aligned} \tag{5.33}$$

since

$$\int_0^1 \Psi(t)\tilde{\Psi}^T(t)dt = I.$$

Applying eqs. (5.29)-(5.33) in the eq. (5.28), we get

$$B^T\Psi(x) = A^T\Theta\tilde{\Psi}(x). \tag{5.34}$$

Multiplying $\Psi^T(x)$ on both sides of eq. (5.34) from the right and integrating with respect to x from 0 to 1, we get

$$B^T P = A^T\Theta, \tag{5.35}$$

where P is a $(2^{M+1}+1)\times(2^{M+1}+1)$ square matrix given by

$$P = \int_0^1 \Psi(x)\Psi^T(x)dx = \begin{pmatrix} P_1 & \\ & P_2 \end{pmatrix}$$

and

$$\int_0^1 \tilde{\Psi}(x)\Psi^T(x) = I.$$

Eq. (5.35) gives a system of $(2^{(M+1)}+1)$ algebraic equations with $2(2^{(M+1)}+1)$ unknowns for A and B given in (5.30) and (5.31).
Again we utilize the following equation

$$g(x, B^T\Psi(x)) = A^T\Psi(x), \tag{5.36}$$

with the collocation points

$$x_s = \frac{s-1}{2^{M+1}} \qquad s = 1, 2, ...2^{M+1}+1. \tag{5.37}$$

Eq. (5.36) gives a system of $(2^{(M+1)}+1)$ algebraic equations with $2(2^{(M+1)}+1)$ unknowns for A and B.

Combining eqs. (5.35) and (5.36), we have a total number of $2(2^{(M+1)}+1)$ algebraic equations with the same number of unknowns for A and B. Solving the system for the unknown coefficients in the vectors A and B, we can obtain the solution $u(x) = B^T\Psi(x)$. Particularly, for $\lambda = 10$, $\beta = 3$ and $\mu = 0.02$, the numerical results obtained by B-spline wavelet method are cited in Table 5.3. Also Table 5.3 cites the comparison of results obtained by the present method and the methods given in ref. [119] (i.e. Adomian's decomposition method, Contraction mapping principle and Shooting method).

5.4 Conclusion

In this present study, the COSMO-RS integral equation has been solved by the Bernstein collocation method, the Haar wavelet method and the Sinc

TABLE 5.3: Comparison of numerical results obtained by B-spline wavelet method with the results of other available methods in ref. [119]

x	B-spline wavelet method $M = 2$	$M = 4$	CMP [119]	SM [119]	ADM [119]
0.0	0.006045	0.006048	0.006079	0.006048	0.006048
0.2	0.018194	0.018193	0.018224	0.018192	0.018192
0.4	0.030424	0.030424	0.030456	0.030424	0.030424
0.6	0.042675	0.042669	0.042701	0.042669	0.042669
0.8	0.054332	0.054368	0.054401	0.054371	0.054371
1.0	0.062030	0.061505	0.061459	0.061458	0.061458

collocation method. All these methods have been applied to solve the Hammerstein integral equation by reducing to a system of algebraic equations. Comparison of numerical results obtained by the three numerical methods, viz. the Bernstein collocation method, the Haar wavelet method, and the Sinc collocation method, have been done in Tables 5.1 and 5.2. It can be clearly observed from the results of Tables 5.1 and 5.2 that the Bernstein collocation method is more powerful and converges more rapidly than other two methods.

In another case, the semi-orthogonal compactly supported linear B-spline wavelets have been applied to solve the nonlinear Hammerstein integral equation which models an adiabatic tubular chemical reactor theory. Using this method, the integral equation has been reduced to a system of algebraic equations. The numerical results obtained by the present method have been compared with the results obtained by Contraction mapping principle, Shooting method, and Adomian's decomposition method and this comparision verifies that the present method gives accurate results with regard to other methods if we increase the value of M.

Chapter 6

Numerical Solution of System of Volterra Integro-Differential Equations

6.1 Introduction

This chapter involves the numerical techniques based on wavelets for solving the Volterra integro-differential equations system. Integral equations occur naturally in many fields of science and engineering [208]. Wavelets theory is a relatively new and emerging area in the field of applied science and engineering. It has been applied in a wide range of engineering disciplines; particularly, wavelets are very successfully used in signal analysis for waveform representations and segmentations, time frequency analysis, and fast algorithms for easy implementation [54]. Wavelets permit the accurate representation of a variety of functions and operators. Moreover wavelets establish a connection with fast numerical algorithms [32]. Wavelets are powerful tools to explore a new direction in solving differential equations and integral equations. In recent years, approximation based on basis functions has been used to estimate the solutions of integral equations, such as orthogonal functions and wavelets. Generally, the sets of piece-wise constant orthogonal functions (e.g., Walsh, Block-Pulse, Haar, etc.), the sets of orthogonal polynomials (e.g., Laguerre, Legendre, Chebyshev, etc.) and the sets of sine-cosine functions in Fourier series have been applied to solve integral equations. One of the most attractive proposals made in the last few years was an idea connected to the application of wavelets as basic functions in the numerical solution of integral equations. The wavelets technique allows the creation of very fast algorithms when compared to the algorithms ordinarily used.

The integral equation has been one of the essential tools for various areas of applied mathematics. Mathematical modeling of real-life problems usually results in functional equations, e.g. partial differential equations, integral and integro-differential equations, stochastic equations and others. Many mathematical formulations of physical phenomena contain integro-differential equations; these equations arise in fluid dynamics, biological models and chemical kinetics. Integro-differential equations arise in many physical processes, such as the glass-forming process [203], nano-hydrodynamics [212], drop wise condensation [198], wind ripple in the desert [45] and biological model [193].

In the past several decades, many effective methods for obtaining approximation or numerical solutions of linear and nonlinear integro-differential equations have been presented. There are various numerical and analytical methods to solve such problems. Nonlinear Volterra integro-differential equations play important roles in the mathematical modeling of many physical and biological phenomena, particularly in such fields as heat transfer, nuclear reactor dynamics, and thermo-elasticity, biological model. Many works have been developed to analyze the numerical methods for solving Volterra integro-differential equations with continuous kernels. In the literature, the Legendre wavelet method (LWM) has been applied to solve the integral equations and integro-differential equations of different forms. Recently, Mohamed and Torky [142] have solved a system of linear Fredholm and Volterra integral equation by applying the Legendre wavelet method. In [201], the learned researchers Venkatesh et al. have applied the Legendre wavelet method to solve a class of nonlinear integro-differential equations. Also the Legendre wavelet method has been applied to solve nonlinear Volterra-Fredholm integral equations by Yousefi and Razzaghi in [217]. Biazer et al. in [39] have solved a system of nonlinear Volterra integro-differential equations by using the Homotopy perturbation method. In [169], a system of linear Volterra integro-differential equations has been solved by using the Variational iteration method by Nadjafi and Tamamgar. Maleknejad et al. [128] have solved a system of high order linear Volterra integro-differential equations by applying the Bernstein operational matrix. Numerical methods for solving Fredholm integral equations have been presented by learned researchers Saha Ray and Sahu in [176]. Linear semi-orthogonal B-spline wavelets have been applied to solve integral equations and systems in [176, 181, 180].

Also, singular integro-differential equations are very difficult to solve analytically as well as numerically. Many researchers have put their interest in numerical methods for solving weakly singular Volterra integro differential equations. In literature, weakly singular integral equations have been solved by many numerical methods, viz. modified Euler method [122], Haar wavelet method [192], wavelet Galerkin method [133], wavelet interpolation method [6] and Chebyshev wavelet method [196]. Homotopy method [171] has been applied to solve singular nonlinear integro-differential equations. A system of nonlinear weakly singular Volterra integro-differential equations has been solved by extrapolation method [77] and quadrature method [151]. Also, the fractional integro-differential equations with weakly singular kernels have been solved by collocation method [228] and the system has been solved by Chebyshev wavelet method [84]. The learned researcher Razlighi et al. have applied the Newton product method to solve the nonlinear singular Volterra integral equations system [18] and also the system of nonlinear singular Volterra integro-differential equations [17] with a particular type of singularity that has been overcome by suitable substitution. In [184], Legendre multi-wavelets have been applied to solve integro-differential forms of Lane-Emden equations. The authors in [182, 185] have solved a system of nonlinear Volterra integro-differential equations by the Legendre wavelet method.

In this chapter, we have applied Legendre wavelets and Bernoulli wavelets to solve a Volterra integro-differential equations system. The properties of Legendre wavelets and Bernoulli wavelets are presented in Chapter 2. By applying this method, the integro-differential equations reduce to systems of algebraic equations and these systems again are solved numerically by any numerical method. In Section 6.2 of this chapter, the system of nonlinear Volterra integro-differential equations has been solved by Legendre wavelet method and in Section 6.3, a nonlinear Volterra weakly singular integro-differential equations system has been solved by Bernoulli wavelet method.

6.2 Legendre wavelet method for solving a system of nonlinear Volterra integro-differential equations

In this section, we consider the system of nonlinear Volterra integro-differential equations of the following form

$$y_i^{(p)}(x) = G_i\left(x, \mathbf{Y}(x)\right) + \sum_{j-1}^{l} \int_0^x k_{i,j}(x,t) F_{i,j}\left(t, \mathbf{Y}(t)\right) dt, \quad i = 1, 2, ..., l. \tag{6.1}$$

with initial conditions $y_i^{(s)}(0) = \beta_{i,s}, \quad s = 0, 1, ..., p-1, \;\; i = 1, 2, ..., l$, where

$$G_i\left(x, \mathbf{Y}(x)\right) = G_i\left(x, y_1(x), y_1^{(1)}(x), ..., y_1^{(p)}(x), ..., y_l(x), y_l^{(1)}(x), ..., y_l^{(p)}(x)\right),$$
$$F_{i,j}\left(t, \mathbf{Y}(t)\right) = F_{i,j}\left(t, y_1(t), y_1^{(1)}(t), ..., y_1^{(p)}(t), ..., y_l(t), y_l^{(1)}(t), ..., y_l^{(p)}(t)\right),$$

and $k_{i,j}(x,t)$ are the kernel functions for $i, j = 1, 2, ..., l$ and $y_i^{(p)}(x)$ is the pth order derivative of $y_i(x)$ for $i = 1, 2, ..., l$.

We have to solve the system of nonlinear Volterra integro-differential equations given in eq. (6.1) by Legendre wavelet method which has already been discussed in Chapter 2. First, we approximate the unknown functions by Legendre wavelet method as

$$y_i(x) = A_i^T \Psi(x), \quad i = 1, 2, ..., l, \tag{6.2}$$

where A_i^T is similar to C^T defined in eq. (2.32) and $\Psi(x)$ is defined in eq. (2.33) of Chapter 2. Let us define $L \equiv \frac{d^p}{dx^p}$; $L^{-1} \equiv \int_0^x \int_0^x ... \int_0^x (.)\underbrace{dxdx...dx}_{p-times}$.

We can also approximate the derivative and integration of $y_i(x)$ $\;i = 1, 2, ..., l$ by operational matrices of Legendre wavelets as

$$y_i^{(1)}(x) = A_i^T \Psi^{(1)}(x) = A_i^T D\Psi(x),$$
$$y_i^{(2)}(x) = A_i^T \Psi^{(2)}(x) = A_i^T D^2\Psi(x),$$
$$\vdots$$
$$y_i^{(p)}(x) = A_i^T \Psi^{(p)}(x) = A_i^T D^p\Psi(x), \quad i = 1, 2, ..., l.$$

similarly,

$$\int_0^x y_i(x)dx = A_i^T \int_0^x \Psi(x)dx = A_i^T Q\Psi(x),$$
$$\int_0^x \int_0^x y_i(x)dxdx = A_i^T \int_0^x \int_0^x \Psi(x)dxdx = A_i^T Q^2\Psi(x),$$
$$\vdots$$
$$\int_0^x ... \int_0^x y_i(x)\underbrace{dxdx...dx}_{p-times} = A_i^T \int_0^x ... \int_0^x \Psi(x)\underbrace{dxdx...dx}_{p-times} = A_i^T Q^p\Psi(x),$$

$i = 1, 2, ..., l,$

where D and Q are defined in eqs. (2.38) and (2.39) of Chapter 2, respectively.

Now substituting eq. (6.2) in eq. (6.1) and then operating L^{-1} both sides, we have

$$\begin{aligned} L^{-1}L\left[A_i^T\Psi(x)\right] =& L^{-1}\left[G_i\left(x,\Omega(x)\right)\right] + \\ & L^{-1}\left[\sum_{j=1}^{l}\int_0^x k_{i,j}(x,t)F_{i,j}\left(t,\Omega(t)\right)dt\right], \\ & i=1,2,...,l, \end{aligned} \tag{6.3}$$

where

$$\Omega(x) = \big(x, A_1^T\Psi(x), A_1^TD^1\Psi(x)..., A_1^TD^p\Psi(x), ..., A_l^T\Psi(x), A_l^TD^1\Psi(x), ..., A_l^TD^p\Psi(x)\big).$$

We set

$$U_i(x) = L^{-1}\left[G_i\left(x,\Omega(x)\right)\right], \quad V_i(x) = L^{-1}\left[\sum_{j=1}^{l}\int_0^x k_{i,j}(x,t)F_{i,j}\left(t,\Omega(t)\right)dt\right].$$

The eq. (6.3) can be reduced as

$$A_i^T\Psi(x) = \sum_{s=0}^{p-1}\beta_{i,s}x^s + U_i(x) + V_i(x), \quad i=1,2,...,l, \tag{6.4}$$

where

$$\beta_{i,s} = y_i^{(s)}(0) = A_i^TD^s\Psi(0), \quad i=1,2,...,l, \quad s=0,1,...,p-1.$$

Utilizing eq. (6.4) by collocation points $x_r = \frac{2r-1}{2^kM}$, $r=1,2,...,(2^{k-1}M)$, we get an algebraic system of $2^{k-1}Ml$ equations along with the same number of unknowns for A_i, $i=1,2,...,l$. Solving this system numerically, we can get the value of unknowns for A_i, $i=1,2,...,l$ and hence obtain the solutions $y_i(x) = A_i^T\Psi(x)$, $i=1,2,...,l$.

6.2.1 Convergence analysis

Theorem 6.2.1. *The series solution* $y(x) \cong \sum_{n=1}^{2^{k-1}}\sum_{m=0}^{M-1} c_{n,m}\psi_{n,m}(x)$ *defined in eq. (2.31) of Chapter 2 using Legendre wavelet method converges to* $y(x)$.

Proof. Let $L^2(\mathbb{R})$ be the Hilbert space and $\psi_{n,m}$ defined in (2.28) of Chapter 2 forms an orthonormal basis.
Let $y(x) = \sum_{i=0}^{M-1} C_{n,i}\psi_{n,i}(x)$ where $C_{n,i} = \langle y(x), \psi_{n,i}(x)\rangle$ for a fixed n.

Let us denote $\psi_{n,i}(x) = \psi(x)$ and let $\alpha_j = \langle y(x), \psi(x)\rangle$.
Now we define the sequence of partial sums S_n of $(\alpha_j \psi(x_j))$; let $\{S_n\}$ and S_m be the partial sums with $n \geq m$. We have to prove S_n is a Cauchy sequence in Hilbert space.
Let $S_n = \sum_{i=1}^{n} \alpha_j \psi(x_j)$.

Now

$$\langle y(x), S_n\rangle = \langle y(x), \sum_{i=1}^{n} \alpha_j \psi(x_j)\rangle = \sum_{j=1}^{n} |\alpha_j|^2.$$

We claim that

$$\|S_n - S_m\|^2 = \sum_{j=m+1}^{n} |\alpha_j|^2, \quad n > m.$$

Now

$$\| \sum_{j=m+1}^{n} \alpha_j \psi(x_j)\|^2 = \langle \sum_{j=m+1}^{n} \alpha_j \psi(x_j), \sum_{j=m+1}^{n} \alpha_j \psi(x_j)\rangle = \sum_{j=m+1}^{n} |\alpha_j|^2, \quad for \quad n > m.$$

Therefore,

$$\| \sum_{j=m+1}^{n} \alpha_j \psi(x_j)\|^2 = \sum_{j=1}^{n} |\alpha_j|^2, \quad for \quad n > m.$$

From Bessel's inequality, we have that $\sum_{j=1}^{n} |\alpha_j|^2$ is convergent and hence

$$\| \sum_{j=m+1}^{n} \alpha_j \psi(x_j)\|^2 \to 0 \quad \text{as} \quad m, n \to \infty$$

$$So, \quad \| \sum_{j=m+1}^{n} \alpha_j \psi(x_j)\| \to 0$$

and $\{S_n\}$ is a Cauchy sequence and it converges to s (say).
We assert that $y(x) = s$.
Now

$$\begin{aligned}\langle s - y(x), \psi(x_j)\rangle =& \langle s, \psi(x_j)\rangle - \langle y(x), \psi(x_j)\rangle \\ =& \langle \lim_{n\to\infty} S_n, \psi(x_j)\rangle - \alpha_j \quad . \\ =& \alpha_j - \alpha_j\end{aligned}$$

This implies

$$\langle s - y(x), \psi(x_j)\rangle = 0.$$

Hence $y(x) = s$ and $\sum_{j=1}^{n} \alpha_j \psi(x_j)$ converges to $y(x)$ as $n \to \infty$ and is proven.

□

TABLE 6.1: Numerical results obtained by Legendre wavelet method with their exact results for Example 6.2.1

x	Numerical solution		Exact solution	
	$y_1(x)$	$y_2(x)$	$y_1(x)$	$y_2(x)$
0	1	-1	1	-1
0.1	1.20517	-1.00517	1.20517	-1.00517
0.2	1.4214	-1.0214	1.4214	-1.0214
0.3	1.64986	-1.04986	1.64986	-1.04986
0.4	1.89182	-1.09182	1.89182	-1.09182
0.5	2.14872	-1.14872	2.14872	-1.14872
0.6	2.42212	-1.22212	2.42212	-1.22212
0.7	2.71375	-1.31375	2.71375	-1.31375
0.8	3.02554	-1.42554	3.02554	-1.42554
0.9	3.3596	-1.5596	3.3596	-1.5596

6.2.2 Illustrative examples

Example 6.2.1. *Consider the following system of Volterra integro-differential equations [39]*

$$y_1''(x) = 1 - \frac{1}{3}x^3 - \frac{1}{2}y_2'^2(x) + \frac{1}{2}\int_0^x \left(y_1^2(t) + y_2^2(t)\right) dt,$$

$$y_2''(x) = -1 + x^2 - xy_1(x) + \frac{1}{4}\int_0^x \left(y_1^2(t) - y_2^2(t)\right) dt,$$

with initial conditions $y_1(0) = 1,\ \ y_1'(0) = 2,\ \ y_2(0) = -1,\ \ y_2'(0) = 0$. *The exact solutions are* $y_1(x) = x + e^x$ *and* $y_2(x) = x - e^x$. *The approximate solutions obtained by Legendre wavelet method for* $M = 8$ *and* $k = 2$ *with their exact solutions have been shown in Table 6.1 and the comparison of absolute errors obtained by Legendre wavelet method and B-spline wavelet method* ($m = 4$) *have been presented in Table 6.2.* L_2 *error for* $y_1(x)$ *obtained by LWM and BWM are 1.09139E-10 and 2.03738E-4, respectively. Similarly,* L_2 *error for* $y_2(x)$ *obtained by LWM and BWM are 1.10249E-10 and 2.07909E-4, respectively. The overall computational time is recorded as 1118.65 seconds for LWM and 30971.20 seconds for BWM for this problem. Clearly, it can be observed that LWM is more accurate than BWM with regard to absolute errors as well as* L_2 *errors. Furthermore, the computational time in LWM is much less than BWM.*

Example 6.2.2. *Consider the following system of Volterra integro-differential equations [39]*

$$y_1'(x) = 1 - \frac{1}{2}y_2'^2(x) + \int_0^x \left((x-t)y_2(t) + y_2(t)y_1(t)\right) dt,$$

$$y_2'(x) = 2x + \int_0^x \left((x-t)y_1(t) - y_2^2(t) + y_1^2(t)\right) dt,$$

TABLE 6.2: Absolute errors obtained by Legendre wavelet method and B-spline wavelet method for Example 6.2.1

x	Legendre wavelet method		B-spline wavelet method	
	$y_1(x)$	$y_2(x)$	$y_1(x)$	$y_2(x)$
0	5.69869E-11	5.71666E-11	1.22179E-4	1.22177E-4
0.1	5.07594E-13	4.49196E-13	4.31693E-5	4.29516E-5
0.2	1.7697E-13	3.01981E-13	1.65916E-5	1.76404E-5
0.3	1.0747E-13	3.05089E-13	4.85740E-5	5.16821E-5
0.4	7.38298E-13	4.65183E-13	4.19752E-5	4.96080E-5
0.5	9.2065E-11	9.42553E-11	1.68426E-5	4.02196E-8
0.6	2.81153E-12	8.09797E-13	3.66785E-5	6.43960E-5
0.7	4.16023E-12	3.58158E-13	4.69159E-5	8.14876E-5
0.8	6.79279E-12	2.73115E-13	1.58263E-6	3.49124E-5
0.9	1.07501E-11	1.12377E-12	1.28407E-4	9.48954E-5

TABLE 6.3: Numerical results obtained by Legendre wavelet method with their exact results for Example 6.2.2

x	Numerical solution		Exact solution	
	$y_1(x)$	$y_2(x)$	$y_1(x)$	$y_2(x)$
0	-1.69785E-5	0.999932	0	1
0.1	0.100168	1.00501	0.100167	1.005
0.2	0.201334	1.02007	0.201336	1.02007
0.3	0.30458	1.04534	0.30452	1.04534
0.4	0.410754	1.08108	0.410752	1.08107
0.5	0.521029	1.12754	0.521095	1.12763
0.6	0.636648	1.18548	0.636654	1.18547
0.7	0.758569	1.25517	0.758584	1.25517
0.8	0.88809	1.33743	0.888106	1.33743
0.9	1.02651	1.43310	1.02652	1.43309

with the initial conditions $y_1(0) = 0$, $y_2(0) = 1$. *The exact solutions are* $y_1(x) = \sinh x$ *and* $y_2(x) = \cosh x$. *The approximate solutions obtained by Legendre wavelet method for* $M = 4$ *and* $k = 2$ *with their exact solutions have been cited in Table 6.3 and the comparison of absolute errors obtained by Legendre wavelet method and B-spline wavelet method* ($m = 2$) *have been presented in Table 6.4.* L_2 *error for* $y_1(x)$ *obtained by LWM and BWM are 7.25142E-5 and 1.32592E-3, respectively. Similarly,* L_2 *error for* $y_2(x)$ *obtained by LWM and BWM are 1.10378E-4 and 2.73105E-3, respectively. The overall computational time is recorded as 424.151 seconds for LWM and 1796.63 seconds for BWM for this problem. Clearly, it can be observed that LWM is more accurate than BWM with regard to absolute errors as well as* L_2 *errors. Furthermore, the computational time in LWM is much less than BWM.*

TABLE 6.4: Absolute errors obtained by Legendre wavelet method and B-spline wavelet method for Example 6.2.2

x	Legendre wavelet method		B-spline wavelet method	
	$y_1(x)$	$y_2(x)$	$y_1(x)$	$y_2(x)$
0	1.69785E-5	6.82587E-5	2.08424E-9	1.98123E-3
0.1	9.28988E-7	1.03030E-5	2.65640E-4	6.34799E-4
0.2	1.57109E-6	1.49865E-6	2.84298E-4	2.57898E-4
0.3	2.17615E-6	1.31948E-6	2.11215E-4	7.05019E-4
0.4	1.24928E-6	1.10079E-5	1.96044E-4	6.67974E-4
0.5	6.63413E-5	8.30807E-5	3.93187E-4	5.88992E-5
0.6	5.12473E-6	1.38480E-5	2.39372E-5	6.76305E-4
0.7	1.48949E-5	8.59612E-7	1.85522E-4	7.83386E-4
0.8	1.64184E-5	9.67972E-7	1.23009E-4	2.82383E-4
0.9	6.44674E-6	1.40323E-5	1.14844E-3	9.83028E-4

Example 6.2.3. *Consider the following system of Volterra integro-differential equations [208]*

$$y_1''(x) = \cosh x - \frac{1}{2}\sinh^2 x - \frac{1}{6}x^4 - \frac{1}{2}x^2 + \int_0^x \left((x-t)y_1^2(t) + (x-t)y_2^2(t)\right) dt,$$

$$y_2''(x) = -(1+4x)\cosh x + 8\sinh x - 4x + \int_0^x \left((x-t)y_1^2(t) - (x-t)y_2^2(t)\right) dt,$$

with the initial conditions $y_1(0) = 1, \quad y_1'(0) = 1$ *and* $y_2(0) = -1 \quad y_2'(0) = 1$. *The exact solutions are* $y_1(x) = x + \cosh x$ *and* $y_2(x) = x - \cosh x$. *The approximate solutions obtained by Legendre wavelet method for* $M = 8$ *and* $k = 2$ *with their exact solutions have been shown in Table 6.5 and the comparison of absolute errors obtained by Legendre wavelet method and B-spline wavelet method* ($m = 4$) *have been presented in Table 6.6.* L_2 *error for* $y_1(x)$ *obtained by LWM and BWM are 7.48238E-11 and 1.65465E-4, respectively. Similarly,* L_2 *error for* $y_2(x)$ *obtained by LWM and BWM are 7.47871E-11 and 1.65577E-4, respectively. The overall computational time is recorded as 1512.60 seconds for LWM and 40879.60 seconds for BWM for this problem. Clearly, it can be observed that LWM is more accurate than BWM with regard to absolute errors as well as* L_2 *errors. Furthermore, the computational time in LWM is much less than BWM.*

6.3 Bernoulli wavelet method for nonlinear Volterra weakly singular integro-differential equations system

In this section, we consider the system of nonlinear weakly singular Volterra integro-differential equations which has the following form:

TABLE 6.5: Numerical results obtained by Legendre wavelet method with their exact results for Example 6.2.3

x	Numerical solution		Exact solution	
	$y_1(x)$	$y_2(x)$	$y_1(x)$	$y_2(x)$
0	1	-1	1	-1
0.1	1.105	-0.905004	1.105	-0.905004
0.2	1.22007	-0.820067	1.22007	-0.820067
0.3	1.34534	-0.745339	1.34534	-0.745339
0.4	1.48107	-0.681072	1.48107	-0.681072
0.5	1.62763	-0.627626	1.62763	-0.627626
0.6	1.78547	-0.585465	1.78547	-0.585465
0.7	1.95517	-0.555169	1.95517	-0.555169
0.8	2.13743	-0.537435	2.13743	-0.537435
0.9	2.33309	-0.533086	2.33309	-0.533086

TABLE 6.6: Absolute errors obtained by Legendre wavelet method and B-spline wavelet method for Example 6.2.3

x	Legendre wavelet method		B-spline wavelet method	
	$y_1(x)$	$y_2(x)$	$y_1(x)$	$y_2(x)$
0	4.69155E-11	4.69159E-11	1.22177E-4	1.22177E-4
0.1	3.6593E-13	3.65152E-13	3.92678E-5	3.92671E-5
0.2	2.40918E-13	2.43361E-13	1.49084E-5	1.49195E-5
0.3	2.367E-13	2.44027E-13	4.06007E-5	4.06552E-5
0.4	3.85914E-13	3.68039E-13	3.64909E-5	3.66545E-5
0.5	5.82778E-11	5.82324E-11	3.42951E-7	2.43668E-8
0.6	5.1581E-13	4.55191E-13	4.04541E-5	4.11247E-5
0.7	2.05169E-13	3.06866E-13	4.86185E-5	4.96506E-5
0.8	1.47882E-13	3.16414E-13	1.83757E-5	1.97573E-5
0.9	7.10543E-13	4.46421E-13	5.80461E-5	5.64004E-5

$$\mathbf{Y}'(x) = \mathbf{F}(x, \mathbf{Y}(x)) + \int_0^x \mathbf{G}(x, t, \mathbf{Y}(t))dt, \quad 0 \le x \le 1 \tag{6.5}$$

with initial conditions $\mathbf{Y}_0(x)$, where,

$$\begin{aligned}
&\mathbf{Y}(x) = (y_1(x), y_2(x), ..., y_d(x))^T, \\
&\mathbf{Y}_0(x) = (y_1(0), y_2(0), ..., y_d(0))^T, \\
&\mathbf{Y}'(x) = (y_1'(x), y_2'(x), ..., y_d'(x))^T, \\
&\mathbf{F}(x, \mathbf{Y}(x)) = (f_1(x, \mathbf{Y}(x)), f_2(x, \mathbf{Y}(x)), ..., f_d(x, \mathbf{Y}(x)))^T, \\
&\mathbf{G}(x, t, \mathbf{Y}(t)) = (G_1(x, t, \mathbf{Y}(t)), G_2(x, t, \mathbf{Y}(t)), ..., G_d(x, t, \mathbf{Y}(t)))^T, \quad d \in \mathbb{N}.
\end{aligned}$$

Here each $G_i(x, t, \mathbf{Y}(t))$ can be written as

$$G_i(x,t,\mathbf{Y}(t)) = K_i(x,t)\tilde{G}_i(x,t,\mathbf{Y}(t)), \quad i = 1, 2, ..., d. \tag{6.6}$$

Also, each element of the vectors $\mathbf{F}(x, \mathbf{Y}(x))$, $\mathbf{G}(x, t, \mathbf{Y}(t))$ is a known function and each element of the vector $\mathbf{Y}(x)$ is an unknown function that has to be determined. $K_i(x,t)$, $i = 1, 2, ..., d$ are called kernel functions which are weakly singular of the form $\frac{\tilde{K}(x,t)}{\sqrt[\alpha]{|x-t|}}$ and $\alpha \in (0,1)$ and $\tilde{G}_i(x,t,\mathbf{Y}(t))$, $i = 1, 2, ..., d$ are nonlinear terms. In order to overcome the singularity, we modify the kernel functions as (see the ref. [133])

$$K_i(x,t) = \begin{cases} \frac{\tilde{K}(x,t)}{\sqrt[\alpha]{|x-t|}}, & x \neq t, \\ 0, & x = t. \end{cases}$$

For solving the nonlinear Volterra weakly singular integro-differential equations system given in equation (6.5), first, integrate eq. (6.5) on both sides with respect to x from 0 to x, we get

$$\mathbf{Y}(x) = \mathbf{Y}_0 + \int_0^x \mathbf{F}(x, \mathbf{Y}(x))dx + \int_0^x \int_0^x \mathbf{G}(x,t,\mathbf{Y}(t))dtdx \tag{6.7}$$

Eq. (6.7) has the form

$$\begin{pmatrix} y_1(x) \\ y_2(x) \\ \cdot \\ \cdot \\ \cdot \\ y_d(x) \end{pmatrix} = \begin{pmatrix} y_1(0) \\ y_2(0) \\ \cdot \\ \cdot \\ \cdot \\ y_d(0) \end{pmatrix} + \begin{pmatrix} \int_0^x f_1(x, y_1(x), ..., y_d(x))dx \\ \int_0^x f_2(x, y_1(x), ..., y_d(x))dx \\ \cdot \\ \cdot \\ \cdot \\ \int_0^x f_d(x, y_1(x), ..., y_d(x))dx \end{pmatrix}$$

$$+ \begin{pmatrix} \int_0^x \int_0^x G_1(x,t,y_1(t),...,y_d(t))dtdx \\ \int_0^x \int_0^x G_2(x,t,y_1(t),...,y_d(t))dtdx \\ \cdot \\ \cdot \\ \cdot \\ \int_0^x \int_0^x G_d(x,t,y_1(t),...,y_d(t))dtdx \end{pmatrix} \tag{6.8}$$

In simplified form, eq. (6.8) can be expressed as

$$y_i(x) = y_i(0) + \int_0^x f_i(x, y_1(x), ..., y_d(x))dx + \int_0^x \int_0^x G_i(x,t,y_1(t),...,y_d(t))dtdx, \quad i = 1, 2, ..., d. \tag{6.9}$$

We approximate the unknown functions $y_i(x), \quad i = 1, 2, ..., d$ by Bernoulli wavelet method as

$$y_i(x) = \sum_{n=1}^{2^{k-1}} \sum_{m=0}^{M-1} c_{i,n,m}\psi_{n,m}(x) = C_i^T \Psi(x), \quad i = 1, 2, ..., d. \tag{6.10}$$

Eq. (6.9) can be written as

$$\begin{aligned} C_i^T \Psi(x) =& y_i(0) + \int_0^x f_i(x, C_1^T \Psi(x), ..., C_d^T \Psi(x))dx \\ &+ \int_0^x \int_0^x G_i(x, t, C_1^T \Psi(t), ..., C_d^T \Psi(t))dtdx, \quad i = 1, 2, ..., d. \end{aligned} \tag{6.11}$$

Putting the collocation points $x_l = \frac{2l-1}{M2^k}, \quad l = 1, 2, ..., M2^{k-1}$ in eq. (6.11), we have

$$\begin{aligned} C_i^T \Psi(x_l) =& y_i(0) + \int_0^{x_l} f_i(x, C_1^T \Psi(x), ..., C_d^T \Psi(x))dx \\ &+ \int_0^{x_l} \int_0^x G_i(x, t, C_1^T \Psi(t), ..., C_d^T \Psi(t))dtdx, \quad i = 1, 2, ..., d. \end{aligned} \tag{6.12}$$

Eq. (6.12) gives a system of $dM2^{k-1}$ number of nonlinear algebraic equations with the same number of unknowns for coefficient vectors $C_i, \quad i = 1, 2, ..., d$. Using any mathematical symbolic software this nonlinear algebraic system can be solved numerically by the Newton method with an appropriate initial guess. Hence we can get the values of unknown vectors $C_i, \quad i = 1, 2, ..., d$ and then obtain the solution $y_i(x), \quad i = 1, 2, ..., d$ from eq. (6.10).

6.3.1 Convergence analysis

Theorem 6.3.1. *If $f(x) \in L^2(\mathbb{R})$ is a continuous function defined on $[0, 1]$ and $|f(x)| \leq \tilde{M}$, then the Bernoulli wavelets expansion of $f(x)$ defined in eq. (2.62) of Chapter 2 converges uniformly and also*

$$|c_{n,m}| < \tilde{M} \frac{A}{2^{\frac{k-1}{2}}} \frac{16m!}{(2\pi)^{m+1}},$$

where $A = \frac{1}{\sqrt{\frac{(-1)^{m-1}(m!)^2}{(2m)!}\alpha_{2m}}}$.

Proof. Any function $f(x) \in L^2[0, 1]$ can be expressed by Bernoulli wavelets as

$$f(x) = \sum_{n=1}^{2^{k-1}} \sum_{m=0}^{M-1} c_{n,m}\psi_{n,m}(x),$$

where the coefficients $c_{n,m}$ can be determined as

$$c_{n,m} = \langle f(x), \psi_{n,m}(x) \rangle$$

Now for $n > 0$, $m > 0$,

$$\begin{aligned} c_{n,m} &= \langle f(x), \psi_{n,m}(x) \rangle \\ &= \int_0^1 f(x)\psi_{n,m}(x)dx \\ &= \int_{I_{nk}} f(x)\psi_{n,m}(x)dx \\ &= 2^{\frac{k-1}{2}} A \int_{I_{nk}} f(x)\beta_m(2^{k-1}x - n + 1)dx \end{aligned}$$

where $I_{nk} = [\frac{n-1}{2^{k-1}}, \frac{n}{2^{k-1}})$.
Now, changing the variable $2^{k-1}x - n + 1 = t$, we have

$$c_{n,m} = \frac{1}{2^{\frac{k-1}{2}}} A \int_0^1 f\left(\frac{t+n-1}{2^{k-1}}\right) \beta_m(t)dt.$$

Thus,

$$\begin{aligned} |c_{n,m}| &\leq \frac{A}{2^{\frac{k-1}{2}}} \int_0^1 \left| f\left(\frac{t+n-1}{2^{k-1}}\right)\right| |\beta_m(t)|dt \\ &\leq \frac{A}{2^{\frac{k-1}{2}}} \tilde{M} \int_0^1 |\beta_m(t)|dt \\ &< \frac{A}{2^{\frac{k-1}{2}}} \tilde{M} \frac{16m!}{(2\pi)^{m+1}}, \text{ using the property of Bernoulli polynomials.} \end{aligned}$$

This means that the series $\sum_{n=1}^{2^{k-1}} \sum_{m=0}^{M-1} c_{n,m}$ is absolutely convergent and hence the series

$$\sum_{n=1}^{2^{k-1}} \sum_{m=0}^{M-1} c_{n,m}\psi_{n,m}(x)$$

is uniformly convergent [115]. □

6.3.2 Illustrative examples

Example 6.3.1. *Let us consider the system of nonlinear weakly singular Volterra integro-differential equations*

$$\begin{aligned} y_1'(x) &= f_1(x) - \frac{1}{2}y_2^2(x) + \int_0^x \frac{1}{\sqrt{x-t}} \left((x-t)y_2(t) + y_1(t)y_2(t)\right) dt, \\ y_2'(x) &= f_2(x) + \int_0^x \frac{1}{\sqrt{x-t}} \left((x-t)y_1(t) - y_2^2(t) + y_1^2(t)\right) dt, \end{aligned}$$

TABLE 6.7: Numerical results and absolute errors for Example 6.3.1

x	$y_1(x)$			$y_2(x)$		
	BWM	Exact	Absolute errors	BWM	Exact	Absolute errors
0.1	0.1	0.1	8.04E-13	1.1	1.1	1.61E-11
0.2	0.2	0.2	5.48E-12	1.2	1.2	8.07E-11
0.3	0.3	0.3	2.16E-11	1.3	1.3	2.05E-10
0.4	0.4	0.4	5.79E-11	1.4	1.4	3.95E-10
0.5	0.5	0.5	1.25E-10	1.5	1.5	6.56E-10
0.6	0.6	0.6	2.33E-10	1.6	1.6	9.68E-10
0.7	0.7	0.7	3.94E-10	1.7	1.7	1.33E-9
0.8	0.8	0.8	6.18E-10	1.8	1.8	1.70E-9
0.9	0.9	0.9	9.15E-10	1.9	1.9	2.07E-9
1.0	1.0	1.0	1.30E-9	2.0	2.0	2.40E-9

with initial conditions $y_1(0) = 0$, $y_2(0) = 1$, *and*

$$f_1(x) = 1 + \frac{1}{2}(1+x)^2 - \frac{2}{3}x^{2/3}(3+2x),$$
$$f_2(x) = 1 - \frac{2}{15}\sqrt{x}(2x^2 - 20x - 15).$$

The exact solutions of this problem are given as $y_1(x) = x$ *and* $y_2(x) = x + 1$. *This problem has been solved by Bernoulli wavelet method (BWM) for* $M = 4$, $k = 2$, *which reduces the integral equation to a system of algebraic equations that has been solved by Newton method. The numerical results and absolute errors for Example 6.3.1 have been provided in Table 6.7.*

Example 6.3.2. *Let us consider the system of nonlinear weakly singular Volterra integro-differential equations*

$$y_1'(x) = f_1(x) + y_1(x) + \int_0^x \frac{x^2 t}{\sqrt{x-t}} y_2^2(t)dt,$$
$$y_2'(x) = f_2(x) + y_2(x) + \int_0^x \frac{xt}{\sqrt[3]{x-t}} y_1^2(t)dt,$$

with initial conditions $y_1(0) = 0$, $y_2(0) = 1$, *and*

$$f_1(x) = \frac{1}{30\sqrt{x}}(15 - 30x - 40x^4 - 32x^5),$$
$$f_2(x) = -0.675x^{11/3} - \frac{1+2x}{2\sqrt{1+x}}.$$

The exact solutions of this problem are given as $y_1(x) = \sqrt{x}$ *and* $y_2(x) = \sqrt{1+x}$. *This problem has been solved by Bernoulli wavelet method (BWM) for* $M = 4$, $k = 2$, *which reduces the integral equation to a system of algebraic equations that has been solved by Newton method. The numerical results and absolute errors for Example 6.3.2 have been presented in Table 6.8.*

TABLE 6.8: Numerical results and absolute errors for Example 6.3.2

x	$y_1(x)$			$y_2(x)$		
	BWM	Exact	Absolute errors	BWM	Exact	Absolute errors
0.1	0.314381	0.316228	0.0018463	1.0488	1.04881	4.09E-6
0.2	0.449175	0.447214	0.0019613	1.09545	1.09545	1.82E-6
0.3	0.549819	0.547723	0.0020965	1.14018	1.14018	5.75E-6
0.4	0.633161	0.632456	0.0007050	1.18323	1.18322	1.04E-5
0.5	0.709594	0.707107	0.0024876	1.22479	1.22474	4.45E-5
0.6	0.777067	0.774597	0.0024705	1.26502	1.26491	1.10E-4
0.7	0.839433	0.836666	0.0027728	1.30407	1.30384	2.26E-4
0.8	0.897514	0.894427	0.0030869	1.34206	1.34164	4.22E-4
0.9	0.952134	0.948683	0.0034507	1.37915	1.37840	7.41E-4
1.0	1.00412	1.000000	0.0041154	1.41545	1.41421	1.24E-3

Example 6.3.3. *As the third example, we consider the following nonlinear singular Volterra integro-differential equations of the form [17]*

$$y_1'(x) = 1 - \frac{\pi}{16} y_1^2(x)(5y_2(x) + 1) + \int_0^x \frac{y_1^2(t)y_2(t)}{\sqrt{t(x-t)}} dt,$$

$$y_2'(x) = 1 - \pi y_2(x) - \frac{3\pi}{4} y_1^2(x) + \int_0^x \frac{y_1^2(x) + y_2^2(x)}{\sqrt{t(x-t)}} dt,$$

with initial conditions $y_1(0) = 0$ *and* $y_2(0) = 1$. *This problem has exact solutions* $y_1(x) = x$ *and* $y_2(x) = x + 1$. *This problem has been solved by Bernoulli wavelet method (BWM) for* $M = 4$, $k = 2$, *which reduces the integral equation to a system of algebraic equations that has been solved by Newton method. The numerical results, in terms of relative errors, obtained by the present method have been compared with that by Newton-Product method (NPM) [17] and this comparison has been presented in Table 6.9. Relative errors are defined as*

$$Relative \quad error = \left|\frac{y_j(x) - y_j^*(x)}{y_j}\right|$$

where $y_j(x)$ *and* $y_j^*(x)$ *are exact and approximate results, respectively, for* $j = 1, 2$.

6.4 Conclusion

In this study, wavelets techniques are applied to solve a nonlinear Volterra integro-differential equations system. The Legendre wavelet method and the

TABLE 6.9: Comparison of relative errors obtained by BWM and NPM for Example 6.3.3

x	BWM solution		NPM solution [17]	
	$\left\lvert\frac{y_1(x)-y_1^*(x)}{y_1}\right\rvert$	$\left\lvert\frac{y_2(x)-y_2^*(x)}{y_2}\right\rvert$	$\left\lvert\frac{y_1(x)-y_1^*(x)}{y_1}\right\rvert$	$\left\lvert\frac{y_2(x)-y_2^*(x)}{y_2}\right\rvert$
0.002	0.000083673	1.66319E-7	0.000356	0.000017
0.003	0.0000542335	1.62357E-7	0.000818	0.000058
0.004	0.0000395324	1.58471E-7	0.001483	0.000139
0.005	0.0000307263	1.54663E-7	0.002364	0.000275
0.006	0.0000248674	1.50934E-7	0.003474	0.000483
0.007	0.0000206925	1.47283E-7	0.004825	0.000779
0.008	0.0000175698	1.43713E-7	0.006434	0.001181
0.009	0.0000151484	1.40224E-7	0.008314	0.001708
0.01	0.0000132177	1.36817E-7	0.010482	0.002382

Bernoulli wavelet method have been applied to solve a system of nonlinear Volterra integro-differential equations. These methods reduce the integro-differential equations system to a system of nonlinear algebraic equations and that algebraic system has been solved by Newton method. In Section 6.2, the obtained numerical results highly agree with the exact results. Additionally, the results obtained by the present method have been compared with that by the B-spline wavelet method. Since, the same solution method has been implemented on both the methods, from the tables it manifests that the present method solutions are better than other method solutions. Since, by this time, the system of nonlinear weakly singular Volterra integro-differential equations has not been solved by any researchers, in Section 6.3, the authors have gained interest to solve this type of integral equations system and consequently solved two Examples 6.3.1 and 6.3.2. Additionally, Example 6.3.3 has been solved for verifying the applicability of the present method. Furthermore, we can get very less absolute error by increasing the order of Bernoulli polynomials. The illustrative examples have been included to demonstrate the validity and applicability of the proposed techniques. These examples also exhibit the accuracy and efficiency of the present method.

Chapter 7

Numerical Solutions of Volterra Integro-Differential Equation Form of Lane-Emden Type Differential Equations

7.1 Introduction

Today, the studies of singular initial value problems in the second order ordinary differential equations (ODEs) of Lane-Emden type have wide applications in mathematical physics and astrophysics [49, 57, 162, 207, 222, 205, 209, 99, 11]. The well-known Lane-Emden equation has been used to model several phenomena in mathematical physics and astrophysics such as the theory of stellar structure, the thermal behavior of a spherical cloud of gas, isothermal gas spheres, the theory of thermionic currents, and the modeling of clusters of galaxies [49, 57, 162]. A substantial amount of work has been done on these types of problems for various structures. The singular behavior that occurs at $x = 0$ is the main difficulty of the Lane-Emden equation. Our main work is to establish a Volterra integro-differential equation equivalent to the Lane-Emden equations of first or second kind. The newly established Volterra integro-differential equation will be solved by using the orthogonal wavelets. Many researchers started using various wavelets for analyzing problems of high computational complexity. It is proven that wavelets are powerful tools to explore a new direction in solving differential equations and integral equations. In this chapter, we have applied Legendre multi-wavelets and Chebyshev wavelets to solve Volterra type integro-differential equations. In Section 7.2, we establish a Volterra integro-differential equation equivalent to the Lane-Emden equation of first and second kind and the newly established Volterra integro-differential equation will be solved by using the Legendre multi-wavelet method (LMWM). The Legendre multi-wavelet method has been applied to solve the integral equations and integro-differential equations of different forms [125, 137, 217, 38, 229]. In Section 7.3, another type of Lane-Emden equation has been considered and solved by Chebyshev wavelet method. The Chebyshev wavelet method has been applied to solve the integral equations and integro-differential equations of different forms [20, 5, 3, 230]. Abd-Elameed et al. have solved singular differential equations and boundary value problems by using Chebyshev wavelets in [3, 4]. The Chebyshev wavelet method for boundary value problems has been discussed in [7]. Biazar et al. have solved a system of integro-differential equations by Chebyshev wavelet method [40]. Also Lane-Emden type differential equations have been solved by Chebyshev polynomials [58]. Properties and function approximation of Legendre multi-wavelets and Chebyshev wavelets have been discussed in Chapter 2. These wavelet methods convert the Volterra integro-differential equation to a system of algebraic equations in the aid of the Gauss-Legendre rule and that algebraic equations system again can be solved numerically by Newton's method. Illustrative examples have been discussed to demonstrate the validity and applicability of the present methods.

7.2 Legendre multi-wavelet method for Volterra integro-differential equation form of Lane-Emden equation

In this section, we discuss the Lane-Emden equation of first kind [118, 207, 205, 210, 209] of the form

$$y'' + \frac{\kappa}{x}y' + y^m = 0, \qquad y(0) = 1, \quad y'(0) = 0, \qquad \kappa > 1 \tag{7.1}$$

and Lane-Emden equation of second kind [98, 153, 216] of the form

$$y'' + \frac{\kappa}{x} y' + e^y = 0, \qquad y(0) = 1, \quad y'(0) = 0, \qquad \kappa \geq 1 \tag{7.2}$$

where κ is the shape factor.

Equation (7.1) is a basic equation in the theory of stellar structure [49]. It is used in astrophysics for computing the structure of interiors of polytropic stars. This equation describes the temperature variation of a spherical gas cloud under the mutual attraction of its molecules and subject to the laws of thermodynamics [205]. In addition, the Lane-Emden equation of the first kind appears also in other contexts such as in the case of radiatively cooling, self-gravitating gas clouds, in the mean-field treatment of a phase transition in critical adsorption and in the modeling of clusters of galaxies.

Equation (7.2) is the Lane-Emden equation of the second kind that models the non-dimensional density distribution $y(x)$ in an isothermal gas sphere [197]. In the study of stellar structures one considers the star as a gaseous sphere in thermodynamic and hydrostatic equilibrium for a certain equation of state [66].

7.2.1 Volterra integro-differential form of the Lane-Emden equation

The generalized form of the Lane-Emden equation is considered as follows

$$y''(x) + \frac{\kappa}{x} y'(x) + f(y) = 0, \qquad y(0) = \alpha, \quad y'(0) = 0, \qquad \kappa \geq 1, \tag{7.3}$$

where $f(y)$ can take any linear or nonlinear forms.
To convert eq. (7.3) into an integro-differential equation, we first set

$$y(x) = \alpha - \frac{1}{\kappa - 1} \int_0^x t \left(1 - \frac{t^{\kappa-1}}{x^{\kappa-1}}\right) f(y(t))dt. \tag{7.4}$$

Differentiating eq. (7.4) twice and using the Leibniz rule, we have

$$y'(x) = -\int_0^x \left(\frac{t^\kappa}{x^\kappa}\right) f(y(t))dt, \tag{7.5}$$

$$y''(x) = -f(y(x)) + \int_0^x \kappa \left(\frac{t^\kappa}{x^{\kappa+1}}\right) f(y(t))dt. \tag{7.6}$$

Multiplying $y'(x)$ by $\frac{\kappa}{x}$ and adding the result to $y''(x)$ gives the generalized Lane-Emden equation (7.3). This shows that the Volterra integro-differential equation equivalent to the generalized Lane-Emden equation (7.3) is given by

$$y'(x) = -\int_0^x \left(\frac{t^\kappa}{x^\kappa}\right) f(y(t))dt, \qquad \kappa \geq 1, \quad y(0) = \alpha. \tag{7.7}$$

7.2.2 Legendre multi-wavelet method for Volterra integro-differential equation form of Lane-Emden equation

Consider the Volterra integro-differential equation given in eq. (7.7) which is the form of Lane-Emden equation defined in eq. (7.3). In order to apply the Legendre multi-wavelets , we first approximate the unknown function $y(x)$ as

$$y(x) = C^T\Psi(x), \tag{7.8}$$

where C is defined similar to eq. (2.43) of Chapter 2.
First, integrating eq. (7.7) and using the initial condition $y(0) = \alpha$, we have

$$y(x) = \alpha - \int_0^x \left[\int_0^z \left(\frac{t^\kappa}{z^\kappa}\right) f(y(t))dt\right] dz, \qquad \kappa \geq 1. \tag{7.9}$$

Then from eqs. (7.8) and (7.9), we have

$$\begin{aligned} C^T\Psi(x) =& \alpha - \int_0^x \left[\int_0^z \left(\frac{t^\kappa}{z^\kappa}\right) f(C^T\Psi(t))dt\right] dz, \qquad \kappa \geq 1 \\ =& \alpha - \int_0^x H(z)dz, \end{aligned} \tag{7.10}$$

where,

$$H(z) = \int_0^z \left(\frac{t^\kappa}{z^\kappa}\right) f(C^T\Psi(t))dt.$$

Now we collocate the eq. (7.10) at the points $x_i = \frac{(2i-1)T}{2^{k+1}(M+1)}$, $i = 1, 2, ..., 2^k(M+1)$ yielding

$$C^T\Psi(x_i) = \alpha - \int_0^{x_i} H(z)dz. \tag{7.11}$$

In order to use the Gaussian integration formula for eq. (7.11), we transfer the interval $[0, x_i]$ into the interval $[-1, 1]$ by means of the transformation

$$\tau = \frac{2}{x_i}z - 1.$$

Eq. (7.11) can be written as

$$C^T\Psi(x_i) = \alpha - \frac{x_i}{2}\int_{-1}^1 H\left(\frac{x_i}{2}(\tau+1)\right) d\tau. \tag{7.12}$$

By using the Gaussian integration formula, we get

$$C^T\Psi(x_i) \cong \alpha - \frac{x_i}{2}\sum_{j=1}^{s} w_j H\left(\frac{x_i}{2}(\tau_j+1)\right), \tag{7.13}$$

where τ_j are s zeros of Legendre polynomials P_{s+1} and w_j are the corresponding weights. The idea behind the above approximation is the exactness of the

Gaussian integration formula for polynomials of degree not exceeding $2s+1$. Eq. (7.13) gives a system of $2^k(M+1)$ nonlinear algebraic equations with the same number of unknowns for coefficient matrix C. Solving this system numerically by Newton's method, we can get the values of unknowns for C and hence we obtain the solution $y(x) = C^T\Psi(x)$.

7.2.3 Convergence analysis

Theorem 7.2.1. *The series solution* $y(x) = \sum_{n=0}^{2^k-1}\sum_{m=0}^{M} c_{n,m}\psi_{n,m}(x)$ *defined in eq.* (2.42) *of Chapter 2 using Legendre multi-wavelet method converges to* $y(x)$.

Proof. Let $L^2(\mathbb{R})$ be the Hilbert space and $\psi_{n,m}$ defined in (2.40) forms an orthonormal basis.
Let $y(x) = \sum_{m=0}^{M} C_{n,m}\psi_{n,m}(x)$ where $C_{n,m} = \langle y(x), \psi_{n,m}(x)\rangle$ for a fixed n.
Let us denote $\psi_{n,m}(x) = \psi(x)$ and let $\alpha_j = \langle y(x), \psi(x)\rangle$.
Now we define the sequence of partial sum $\{S_n\}$ of $(\alpha_j\psi(x_j))$; Let $\{S_n\}$ and $\{S_m\}$ be the partial sums with $n \geq m$. We have to prove $\{S_n\}$ is a Cauchy sequence in Hilbert space.
Let $S_n = \sum_{j=1}^{n}\alpha_j\psi(x_j)$.
Now

$$\langle y(x), S_n\rangle = \langle y(x), \sum_{j=1}^{n}\alpha_j\psi(x_j)\rangle = \sum_{j=1}^{n}|\alpha_j|^2.$$

We claim that

$$\|S_n - S_m\|^2 = \sum_{j=m+1}^{n}|\alpha_j|^2, \quad n > m.$$

Now

$$\|\sum_{j=m+1}^{n}\alpha_j\psi(x_j)\|^2 = \langle\sum_{j=m+1}^{n}\alpha_j\psi(x_j), \sum_{j=m+1}^{n}\alpha_j\psi(x_j)\rangle = \sum_{j=m+1}^{n}|\alpha_j|^2, \quad for \quad n > m.$$

Therefore,

$$\|\sum_{j=m+1}^{n}\alpha_j\psi(x_j)\|^2 = \sum_{j=m+1}^{n}|\alpha_j|^2, \quad for \quad n > m.$$

From Bessel's inequality, we have that $\sum_{j=1}^{\infty}|\alpha_j|^2$ is convergent and hence

$$\|\sum_{j=m+1}^{n}\alpha_j\psi(x_j)\|^2 \to 0 \quad \text{as} \quad n \to \infty$$

So,

$$\|\sum_{j=m+1}^{n}\alpha_j\psi(x_j)\| \to 0$$

and $\{S_n\}$ is a Cauchy sequence and it converges to s (say).
We assert that $y(x) = s$.
Now

$$\begin{aligned}\langle s - y(x), \psi(x_j)\rangle =& \langle s, \psi(x_j)\rangle - \langle y(x), \psi(x_j)\rangle \\ =& \langle \lim_{n\to\infty} S_n, \psi(x_j)\rangle - \alpha_j \\ =& \alpha_j - \alpha_j\end{aligned}$$

This implies

$$\langle s - y(x), \psi(x_j)\rangle = 0$$

Hence $y(x) = s$ and $\sum_{j=1}^{n} \alpha_j \psi(x_j)$ converges to $y(x)$ as $n \to \infty$ and is proven.
□

7.2.4 Illustrative examples

Example 7.2.1. *Consider the generalized form of Lane-Emden equation of first kind*

$$y''(x) + \frac{\kappa}{x} y'(x) + y^m(x) = 0, \qquad \kappa \geq 1, \quad y(0) = 1, \quad y'(0) = 0.$$

This equation can be transformed to the integro-differential form as follows

$$y'(x) = -\int_0^x \left(\frac{t^\kappa}{x^\kappa}\right) y^m(t)dt, \qquad y(0) = 1, \quad \kappa \geq 1.$$

The exact solutions of this problem for $\kappa = 2$ *and* $m = 0, 1, 5$ *respectively are given by*

$$\begin{aligned} y(x) &= 1 - \frac{1}{3!}x^2, \\ y(x) &= \frac{\sin x}{x}, \\ y(x) &= \left(1 + \frac{x^2}{3}\right)^{-\frac{1}{2}}. \end{aligned}$$

The approximate solutions obtained by Legendre multi-wavelet method (LMWM) $(M = 7, k = 1)$ *for shape factor* $\kappa = 2$ *and* $m = 0, 1, 5$ *with their corresponding exact solutions and absolute errors have been shown in Tables 7.1-7.3, respectively.*

Example 7.2.2. *Consider the Lane-Emden equation of second kind*

$$y''(x) + \frac{\kappa}{x} y'(x) + e^{y(x)} = 0, \qquad y(0) = y'(0) = 0, \quad \kappa > 1.$$

TABLE 7.1: Numerical solutions for Example 7.2.1 when $\kappa = 2, \quad m = 0$

x	LMWM solution	Exact solution	Absolute error
0.2	0.993333	0.993333	2.66664E-12
0.4	0.973333	0.973333	2.13333E-11
0.6	0.940000	0.940000	7.20001E-11
0.8	0.893333	0.893333	1.70667E-10
1	0.833333	0.833333	3.33333E-10

TABLE 7.2: Numerical solutions for Example 7.2.1 when $\kappa = 2, \quad m = 1$

x	LMWM solution	Exact solution	Absolute error
0.2	0.993347	0.993347	2.45593E-9
0.4	0.973546	0.973546	5.46664E-10
0.6	0.941071	0.941071	2.45289E-10
0.8	0.896695	0.896695	1.94895E-10
1	0.841471	0.841471	2.45936E-10

TABLE 7.3: Numerical solutions for Example 7.2.1 when $\kappa = 2, \quad m = 5$

x	LMWM solution	Exact solution	Absolute error
0	1	1	2.66055E-9
0.2	0.993399	0.993399	1.07934E-11
0.4	0.974355	0.974355	1.17952E-11
0.6	0.944911	0.944911	1.64531E-11
0.8	0.907841	0.907841	2.17233E-11

TABLE 7.4: Numerical solutions for Example 7.2.2

x	$\kappa = 2$		$\kappa = 3$		$\kappa = 4$	
	LMWM	VIM	LMWM	VIM	LMWM	VIM
0	-5.7433E-11	0	-2.484E-11	0	-1.2637E-11	0
0.2	-0.006653	-0.006653	-0.004992	-0.004992	-0.003994	-0.003994
0.4	-0.026456	-0.026456	-0.019868	-0.019868	-0.015909	-0.015909
0.6	-0.058944	-0.058944	-0.044337	-0.044337	-0.035544	-0.035544
0.8	-0.103386	-0.103386	-0.077935	-0.077935	-0.062578	-0.062578

TABLE 7.5: Numerical solutions for Example 7.2.3

x	LMWM solution	Exact solution	Absolute error
0	1	1	3.95615E-8
0.1	0.990050	0.990050	2.96242E-10
0.2	0.960789	0.960789	3.82808E-10
0.3	0.913931	0.913931	2.95619E-8
0.4	0.852143	0.852143	4.68592E-7
0.5	0.778797	0.778797	3.64064E-6

This equation can be transformed to integro-differential form as follows

$$y'(x) = -\int_0^x \left(\frac{t^\kappa}{x^\kappa}\right) e^{y(t)} dt, \qquad y(0) = 1, \quad \kappa > 1.$$

The approximate solutions obtained by Legendre multi-wavelet method ($M = 7, k = 1$) for shape factor $\kappa = 2, 3, 4$ have been compared with the solutions obtained by variational iteration method (VIM) [209] cited in Table 7.4.

Example 7.2.3. *Next, consider the Lane-Emden type equation given by*

$$y''(x) + \frac{8}{x}y'(x) + (18y(x) + 4y(x)lny(x)) = 0, \qquad y(0) = 1, \quad y'(0) = 0.$$

The Volterra integro-differential form of this equation is given by

$$y'(x) + \int_0^x \frac{t^8}{x^8}(18y(t) + 4y(t)lny(t))dt = 0, \quad y(0) = 1,$$

with exact solution e^{-x^2}. The Legendre multi-wavelets solutions for $M = 7, k = 1$ along with their corresponding exact solutions and absolute errors have been shown in Table 7.5.

Example 7.2.4. *Consider the Lane-Emden type equation given by*

$$y''(x) + \frac{1}{x}y'(x) + (3y^5(x) - y^3(x)) = 0, \qquad y(0) = 1, \quad y'(0) = 0.$$

TABLE 7.6: Numerical solutions for Example 7.2.4

x	LMWM solution	Exact solution	Absolute error
0	1	1	9.41731E-8
0.2	0.980581	0.980581	8.91026E-10
0.4	0.928477	0.928477	1.53517E-9
0.6	0.857493	0.857493	1.16852E-9
0.8	0.780869	0.780869	1.55470E-9

TABLE 7.7: Numerical solutions for Example 7.2.5

x	LMWM solution	Exact solution	Absolute error
0	1.1743E-7	0	1.17430E-7
0.2	-0.078441	-0.078441	1.25003E-9
0.4	-0.296840	-0.296840	1.65908E-7
0.6	-0.614985	-0.614969	1.52712E-5
0.8	-0.989704	-0.989392	3.11348E-4

The Volterra integro-differential form of this equation is given by

$$y'(x) + \int_0^x \frac{t}{x}(3y^5(t) - y^3(t))dt = 0, \quad y(0) = 1,$$

with exact solution $\frac{1}{\sqrt{1+x^2}}$. *The Legendre multi-wavelets solutions for* $M = 7, k = 1$ *along with their corresponding exact solutions and absolute errors have been shown in Table 7.6.*

Example 7.2.5. *Consider the Lane-Emden type equation given by*

$$y''(x) + \frac{2}{x}y'(x) + 4\left(2e^{y(x)} + e^{\frac{y(x)}{2}}\right) = 0, \qquad y(0) = y'(0) = 0.$$

The Volterra integro-differential form of this equation is given by

$$y'(x) + \int_0^x \frac{t^2}{x^2}\left(4\left(2e^{y(t)} + e^{\frac{y(t)}{2}}\right)\right)dt = 0, \quad y(0) = 0,$$

with exact solution $-2ln(1 + x^2)$. *The Legendre multi-wavelets solutions for* $M = 7, k = 1$ *along with their corresponding exact solutions and absolute errors have been shown in Table 7.7.*

Example 7.2.6. *Consider the system of nonlinear Lane-Emden type equations given by*

$$y_1''(x) + \frac{8}{x}y_1'(x) + (18y_1(x) - 4y_1(x)lny_2(x)) = 0,$$
$$y_2''(x) + \frac{4}{x}y_2'(x) + (4y_2(x)lny_1(x) - 10y_2(x)) = 0,$$

TABLE 7.8: Numerical solutions for Example 7.2.6

x	LMWM solution		Exact solution		Absolute error	
	$y_1(x)$	$y_2(x)$	$y_1(x)$	$y_2(x)$	$y_1(x)$	$y_2(x)$
0	1	1	1	1	7.15876E-8	8.44232E-8
0.1	0.99005	1.01005	0.99005	1.01005	5.61584E-10	6.59049E-10
0.2	0.960789	1.04081	0.960789	1.04081	9.69923E-10	3.34747E-10
0.3	0.913931	1.09417	0.913931	1.09417	3.5286E-8	4.47131E-8
0.4	0.852144	1.17351	0.852144	1.17351	6.22823E-7	8.00388E-7
0.5	0.778805	1.28402	0.778801	1.28403	4.48153E-6	7.03964E-6

with initial conditions

$$y_1(0) = 1, \quad y_1'(0) = 0,$$
$$y_2(0) = 1, \quad y_2'(0) = 0.$$

The nonlinear Volterra integro-differential form of the above system is given by

$$y_1'(x) + \int_0^x \frac{t^8}{x^8}(18y_1(t) - 4y_1(t)lny_2(t))dt = 0,$$
$$y_2'(x) + \int_0^x \frac{t^4}{x^4}(4y_2(t)lny_1(t) - 10y_2(t))dt = 0,$$

with initial conditions $y_1(0) = 1, \quad y_2(0) = 1.$
The corresponding exact solutions of this system are

$$y_1(x) = e^{-x^2},$$
$$y_2(x) = e^{x^2}.$$

The approximate solutions obtained by Legendre multi-wavelet method for $M = 7, k = 1$ *along with their corresponding exact solutions and absolute errors have been shown in Table 7.8.*

7.3 Chebyshev wavelet method for Volterra integro-differential equation form of Lane-Emden type equation

In this section, singular initial value problems in the second order ordinary differential equations (ODEs) of Lane-Emden type equation formulated as

$$y''(x) + \frac{\alpha}{x}y'(x) + p(x)y(x) = g(x), \qquad 0 < x \leq 1, \quad \alpha \geq 0, \tag{7.14}$$

with the following initial conditions

$$y(0) = A, \quad y'(0) = B, \tag{7.15}$$

where A and B are constants, $p(x)$ is the analytic function, and $g(x) \in C[0,1]$.

Equation (7.14) has been used to model several phenomena in mathematical physics and astrophysics such as the theory of stellar structure, the thermal behavior of a spherical cloud of gas, isothermal gas spheres, the theory of thermionic currents, and in the modeling of clusters of galaxies [49, 57, 162]. A substantial amount of work has been done on these types of problems for various structures. The singular behavior that occurs at $x = 0$ is the main difficulty of eq. (7.14).

7.3.1 Volterra integro-differential form of the Lane-Emden type differential equation

In this section, we discuss the conversion of eq. (7.14) into the integro-differential equation; first we set

$$y''(x) = u(x). \tag{7.16}$$

Integrating eq. (7.16) two times with respect to x from 0 to x, we have

$$y(x) = A + Bx + \int_0^x \int_0^\theta u(t)dtd\theta = A + Bx + \int_0^x \left(\int_t^x d\theta\right) u(t)dt,$$

$$or \quad y(x) = A + Bx + \int_0^x (x-t)u(t)dt. \tag{7.17}$$

Putting eqs. (7.16)-(7.17) in eq. (7.14), we have

$$u(x) = f(x) - \int_0^x k(x,t)u(t)dt, \tag{7.18}$$

where

$$f(x) = g(x) - B\left(\frac{\alpha}{x}\right) - Ap(x) - Bxp(x)$$

and

$$k(x,t) = \frac{\alpha}{x} + p(x)(x-t).$$

Differentiating eq. (7.18) with respect to x and applying the Leibnitz rule, we have

$$u'(x) + \frac{\alpha}{x}u(x) = f'(x) - \int_0^x \frac{\partial}{\partial x}k(x,t)u(t)dt. \tag{7.19}$$

Eq. (7.19) can be written as

$$u'(x) + \frac{\alpha}{x}u(x) = F(x) - \int_0^x K(x,t)u(t)dt, \qquad u(0) = u_0, \tag{7.20}$$

where $F(x) = f'(x)$ and $K(x,t) = \frac{\partial}{\partial x} k(x,t)$.
Eq. (7.20) is the required volterra integro-differential form of the Lane-Emden type equation (7.14).

7.3.2 Analysis of method

Consider the Volterra integro-differential equation given in eq. (7.20) which is the form of Lane-Emden equation defined in eq. (7.14). In order to apply the Chebyshev wavelets, we first approximate the unknown function $u(x)$ as

$$u(x) = C^T \Psi(x), \tag{7.21}$$

where C is defined similar to eq. (2.50) of Chapter 2.
First, integrate the eq. (7.20) on both sides with respect to x from 0 to x and using the initial condition $u(0) = u_0$, we have

$$u(x) = u_0 - \int_0^x \frac{\alpha}{x} u(x)dx + \int_0^x F(x)dx - \int_0^x \left[\int_0^x K(x,t)u(t)dt \right] dx. \tag{7.22}$$

Then from eqs. (7.21) and (7.22), we have

$$\begin{aligned} C^T \Psi(x) =& u_0 + \int_0^x F(x)dx - \int_0^x \frac{\alpha}{x} C^T \Psi(x)dx \\ & - \int_0^x \left[\int_0^x K(x,t) C^T \Psi(t)dt \right] dx \\ =& G(x) - \int_0^x H_1(x)dx - \int_0^x H_2(x)dx, \end{aligned} \tag{7.23}$$

where

$$\begin{aligned} G(x) =& u_0 + \int_0^x F(x)dx, \\ H_1(x) =& \frac{\alpha}{x} C^T \Psi(x), \\ H_2(x) =& \int_0^x K(x,t) C^T \Psi(t)dt. \end{aligned}$$

Now, we collocate the eq. (7.23) at $2^{k-1}M$ points $x_i = \frac{2i-1}{2^k M}$, $i = 1, 2, ..., 2^{k-1}M$ as

$$C^T \Psi(x_i) = G(x_i) - \int_0^{x_i} H_1(x)dx - \int_0^{x_i} H_2(x)dx. \tag{7.24}$$

In order to use the Gaussian integration formula for eq. (7.24), we transfer the interval $[0, x_i]$ into the interval $[-1, 1]$ by means of the transformation

$$\tau = \frac{2}{x_i} x - 1.$$

Eq. (7.24) can be written as

$$C^T\Psi(x_i) \cong G(x_i) - \frac{x_i}{2}\int_{-1}^{1} H_1\left(\frac{x_i}{2}(\tau+1)\right)d\tau - \frac{x_i}{2}\int_{-1}^{1} H_2\left(\frac{x_i}{2}(\tau+1)\right)d\tau. \tag{7.25}$$

By using the Gaussian quadrature formula, we get

$$C^T\Psi(x_i) = G(x_i) - \frac{x_i}{2}\sum_{j_1=1}^{s_1} w_{j_1}H_1\left(\frac{x_i}{2}(\tau_{j_1}+1)\right) - \frac{x_i}{2}\sum_{j_2=1}^{s_2} w_{j_2}H_2\left(\frac{x_i}{2}(\tau_{j_2}+1)\right), \tag{7.26}$$

where τ_{j_1} and τ_{j_2} are s_1 and s_2 zeros of Legendre polynomials $P_{s_1+1}(\cdot)$ and $P_{s_2+1}(\cdot)$ respectively and w_{j_1} and w_{j_2} are the corresponding weights. The idea behind the above approximation is the exactness of the Gaussian quadrature formula for polynomials of degree not exceeding $2s_1+1$ and $2s_2+1$ respectively. Eq. (7.26) gives a system of $2^{k-1}M$ algebraic equations with the same number of unknowns for coefficient matrix C. Solving this system numerically, we can get the values of unknowns for C and hence we obtain the solution $u(x) = C^T\Psi(x)$. Again we can calculate the actual solution in term of $y(x)$ by eq. (7.17).

7.3.3 Illustrative examples

Example 7.3.1. *Consider the second order singular differential equation of Lane-Emden type [206, 11]*

$$y''(x) + \frac{2}{x}y'(x) - 2(2x^2+3)y(x) = 0, \quad y(0) = 1, \quad y'(0) = 0, \quad 0 < x \leq 1.$$

This equation can be transformed to integro-differential form as follows

$$u'(x) + \frac{2}{x}u(x) = 8x - \int_0^x \left(\frac{-2}{x^2} - 12x^2 + 8xt - 6\right)u(t)dt, \quad u(0) = 2,$$

with the exact solution $y^*(x) = e^{x^2}$. *The approximate solutions obtained by the Chebyshev wavelet method (CWM)($M = 6, \quad k = 4$) have been compared with that of the* 5*-th order Adomian decomposition method (ADM) [206] along with the corresponding exact solutions and cited in Table 7.9. The obtained* L_2 *and* L_∞ *errors have been shown in Table 7.10. In this problem, we get* 48 *number algebraic equations with the same number of unknowns and these equations have been solved numerically by Newton's method.*

TABLE 7.9: Comparison of approximate solutions obtained by CWM and ADM for Example 7.3.1

x	CWM	ADM	Exact
0.1	1.01005	1.01005	1.01005
0.2	1.04081	1.04081	1.04081
0.3	1.09417	1.09417	1.09417
0.4	1.17351	1.17351	1.17351
0.5	1.28403	1.28403	1.28403
0.6	1.43333	1.43333	1.43333
0.7	1.63232	1.63232	1.63232
0.8	1.89648	1.89648	1.89648
0.9	2.24791	2.2479	2.24791

TABLE 7.10: L_2 and L_∞ errors obtained by CWM and ADM for Example 7.3.1

Error	CWM	ADM
L_2	1.05492E-7	5.59398E-6
L_∞	1.03058E-7	5.59068E-6

The error term can be calculated as

$$Absolute \;\; error = |e_i| = |y^*(x_i) - \sum_{n=1}^{2^{k-1}} \sum_{m=0}^{M-1} c_{n,m}\psi_{n,m}(x_i)|, \quad i = 0, 1, ..., s,$$

$$L_2 \;\; error = \sqrt{\sum_{i=0}^{s} |e_i|^2},$$

$$L_\infty \;\; error = \max_{0 \le i \le s} |e_i|,$$

where s is the number of nodes points in the interval $[0, 1]$.

Example 7.3.2. *Consider the second order singular differential equation of Lane-Emden type [206, 11]*

$$y''(x) + \frac{8}{x}y'(x) + xy(x) = x^5 - x^4 + 44x^2 - 30x, \quad y(0) = y'(0) = 0, \quad 0 < x \le 1.$$

This equation can be transformed to integro-differential form as follows

$$u'(x) + \frac{8}{x}u(x) = 5x^4 - 4x^3 + 88x - 30 - \int_0^x \left(\frac{-8}{x^2} + 2x - t\right) u(t)dt, \quad u(0) = 0,$$

with the exact solution $y^*(x) = x^4 - x^3$. The approximate solutions obtained by Chebyshev wavelet method ($M = 6, \quad k = 1$) have been compared with that

TABLE 7.11: Comparison of approximate solutions obtained by CWM and LWM for Example 7.3.2

x	CWM	LWM	Exact	Error in CWM	Error in LWM
0.1	-0.0009	-0.0009	-0.0009	1.4962E-17	1.6E-11
0.2	-0.0064	-0.0064	-0.0064	4.59702E-17	1.5E-11
0.3	-0.0189	-0.0189	-0.0189	6.245E-17	0
0.4	-0.0384	-0.0384	-0.0384	6.245E-17	1.0E-11
0.5	-0.0625	-0.0625	-0.0625	1.04083E-16	4.0E-11
0.6	-0.0864	-0.0864	-0.0864	1.94289E-16	1.0E-11
0.7	-0.1029	-0.1029	-0.1029	3.33067E-16	1.0E-10
0.8	-0.1024	-0.1024	-0.1024	4.30211E-16	2.0E-10
0.9	-0.0729	-0.0729	-0.0729	4.85723E-16	3.3E-10

TABLE 7.12: L_2 and L_∞ errors obtained by CWM and LWM for Example 7.3.2

Error	CWM	LWM
L_2	7.68544E-16	4.01474E-10
L_∞	4.85723E-16	3.3E-10

of the Legendre wavelet method (LWM) ($M = 6, \quad k = 1$) [11] along with the corresponding exact solutions and cited in Table 7.11. The obtained L_2 and L_∞ errors have been shown in Table 7.12. In this present method, we get 6 number algebraic equations with the same number of unknowns and these equations have been solved numerically by Newton's method.

Example 7.3.3. *Consider the second order singular differential equation of Lane-Emden type [206]*

$$y''(x) + \frac{2}{x}y'(x) + y(x) = 6 + 12x + x^2 + x^3, \quad y(0) = y'(0) = 0, \quad 0 < x \leq 1.$$

This equation can be transformed to integro-differential form as follows

$$u'(x) + \frac{2}{x}u(x) = 12 + 2x + 3x^2 - \int_0^x \left(\frac{-2}{x^2} + 1\right) u(t)dt, \quad u(0) = 2,$$

with the exact solution $y^*(x) = x^2 + x^3$. The approximate solutions obtained by Chebyshev wavelet method ($M = 8, \quad k = 1$) have been compared with that of the 5-th order Adomian decomposition method (ADM) [206] along with the corresponding exact solutions and cited in Table 7.13. The obtained L_2 and L_∞ errors have been shown in Table 7.14. In this present method, we get 8 number algebraic equations with the same number of unknowns and these equations have been solved numerically by Newton's method.

TABLE 7.13: Comparison of approximate solutions obtained by CWM and ADM for Example 7.3.3

x	CWM	ADM	Exact
0.1	0.011	0.011	0.011
0.2	0.048	0.048	0.048
0.3	0.117	0.117	0.117
0.4	0.224	0.224	0.224
0.5	0.375	0.375	0.375
0.6	0.576	0.576	0.576
0.7	0.833	0.833	0.833
0.8	1.152	1.152	1.152
0.9	1.539	1.539	1.539

TABLE 7.14: L_2 and L_∞ errors obtained by CWM and ADM for Example 7.3.3

Error	CWM	ADM
L_2	5.66647E-14	1.30900E-12
L_∞	2.02061E-14	1.28586E-12

7.4 Conclusion

In this chapter, Volterra integro-differential equations equivalent to the Lane-Emden type second order singular differential equations have been established. The newly obtained Volterra integro-differential form of Lane-Emden type equations facilitates the computational work and overcomes the difficulty of the singular behavior at $x = 0$. Using these procedures, the integro-differential forms have been reduced to solve a system of algebraic equations. This algebraic equations system has been solved numerically by Newton's method. The illustrative examples have been included to demonstrate the validity and applicability of the present techniques. From the tables, it is clear that the L_2 and L_∞ errors obtained by the present methods are less than other comparative methods. The illustrative examples in Sections 7.2 and 7.3 also exhibit the accuracy and efficiency of the Legendre multi-wavelet method and Chebyshev wavelet method.

Chapter 8

Application of Legendre Spectral Collocation Method for Solving Integro-Differential Equations

8.1 Introduction

In this chapter, the Legendre spectral collocation method has been applied to solve integro-differential equations. The proposed method is based on the Gauss-Legendre points with the basis functions of Lagrange polynomials. The presented method applied to the integral equation reduces to solve the system of algebraic equations. The Legendre spectral collocation method has been applied to solve the Fredholm integro differential-difference equation with variable coefficients and mixed conditions in Section 8.2 and the system of integro-differential equations modeling biological species living together in Section 8.3. Also, the numerical results obtained by the Legendre spectral collocation method have been compared with the results obtained by existing methods. Illustrative examples have been discussed to demonstrate the validity and applicability of the presented method.

8.2 Legendre spectral collocation method for Fredholm integro-differential-difference equation with variable coefficients and mixed conditions

The studies of integro-differential-difference equations have been developed very rapidly and intensively. In recent years, both mathematicians and physicists have devoted considerable effort to the study of numerical solutions of the integro-differential-difference equations. These equations are combinations in which the unknown functions appear under the sign of integration and they also contain the derivatives and functional arguments of the unknown functions. Problems involving these equations arise frequently in many applied areas which include engineering, mechanics, physics, chemistry, astronomy, biology, economics, potential theory, electrostatics, etc. [72, 76, 71, 166, 213, 103, 92].

Many numerical methods have been presented in open literature for solving integro-differential-difference equations and integro-differential equations. The learned researcher Sezer et al. [72] solved integro-differential-difference equations with variable coefficients by using the Taylor matrix method, the Laguerre collocation method (LCM) [76], and the Chebyshev collocation method [71]. In [166], the higher-order linear Fredholm integro-differential-difference equations with variable coefficients have been solved by Legendre polynomials. The Boubaker polynomial [213], Fibonacci collocation method (FCM) [103] and homotopy analysis method (HAM) [92] have been applied to solve Fredholm integro-differential-difference equations with variable coefficients. The numerical methods for solving integro-differential equations with variable coefficients have been discussed in [139, 224]. Also the differential-difference equations with variable coefficients have been solved by many researchers using different numerical methods. The differential-difference equations with variable coefficients have been solved by a hybrid of Legendre and Taylor polynomials [73], and Jacobian elliptic function method [56]. In [75], the delay difference equations with variable coefficients have been solved by Laguerre polynomials. The differential transform method has been applied to solve the differential-difference equations in [13].

In this work, we consider the general form of mth order linear Fredholm integro-differential-difference equation with variable coefficients as

$$\sum_{i=0}^{m} p_i(x)y^{(i)}(x)+\sum_{j=0}^{n} p_j^*(x)y^{(j)}(x-\tau) = g(x)+\int_a^b K(x,t)y(t-\tau)dt, \tau\geq 0,\ n\leq m, \tag{8.1}$$

with mixed condition

$$\sum_{i=0}^{m-1}\left[\alpha_{il}y^{(i)}(a)+\beta_{il}y^{(i)}(b)+\gamma_{il}y^{(i)}(c)\right]=\mu_i,\quad l=0,1,...,m-1. \tag{8.2}$$

where $p_i(x)$, $p_i^*(x)$, $g(x)$, and $K(x,t)$ are known functions, and α_{il}, β_{il} and γ_{il} are appropriate constants. $y(x)$ is the unknown function that has to be determined. The eq. (8.1) along with eq. (8.2) has been solved by Legendre spectral collocation method (LSCM) . The Legendre spectral collocation method has been applied by many authors to solve integral equations and integro-differential equations [200, 187, 51]. In this method, the unknown function $y(x)$ has been approximated by Lagrange polynomial with the Gauss-Legendre points as node points on the interval $[-1,1]$. In general, the integro-differential-difference equation defined over interval $[a,b]$ can be transformed into the interval $[-1,1]$. The present method reduces the integro-differential-difference equation into a system of algebraic equations that can be solved by any numerical method. Particularly, this system has been solved by the Newton method.

8.2.1 Lagrange polynomial and its properties

Any function $y(t)$ defined over $[a,b]$ can be approximated by the Lagrange interpolation polynomial as

$$y(t)=\sum_{k=0}^{M}y_kF_k(t), \tag{8.3}$$

with $y_k=y(t_k)$, where $t_k,\ k=0,1,...,M$ are interpolating points satisfying $a\le t_0<t_1<...<t_{M-1}<t_M\le b$ and

$$F_k(t)=\prod_{j=0,j\neq k}^{M}\left(\frac{t-t_j}{t_k-t_j}\right),$$

also,

$$F_k(t_r)=\delta_{kr}=\begin{cases}1, & k=r\\ 0, & k\neq r.\end{cases}$$

8.2.2 Analysis of Legendre spectral collocation method

In this section, the basic idea of the Legendre spectral collocation method for solving integro-differential-difference equations has been presented. This idea has been proposed in [200, 187, 51]. In this procedure of approximation, the Legendre-Gauss quadrature rule together with Lagrange interpolation polynomials have been used.

In order to use the Legendre spectral collocation method, we consider the Legendre-Gauss points $\{x_s\}_{s=0}^{M}$ i.e., the roots of $L_{M+1}(x) = 0$ where L_{M+1} is the $(M+1)^{th}$ Legendre polynomial.

Consider the integro-differential-difference equations with variable coefficients defined in eq. (8.1) and substituting the approximate unknown function $y(x)$ obtained by Lagrange interpolation polynomial of eq. (8.3), we obtain

$$\begin{aligned}\sum_{i=0}^{m} p_i(x) \sum_{k=0}^{M} y_k F_k^{(i)}(x) + \sum_{j=0}^{n} p_j^*(x) \sum_{k=0}^{M} y_k F_k^{(j)}(x-\tau) \\ = g(x) + \int_{-1}^{1} K(x,t) \sum_{k=0}^{M} y_k F_k(t-\tau) dt. \qquad (8.4)\end{aligned}$$

Let us consider the integral part of eq. (8.4) and we apply the Legendre-Gauss quadrature rule as

$$\int_{-1}^{1} K(x,t) \sum_{k=0}^{M} y_k F_k(t-\tau) dt \approx \sum_{s_1=0}^{M} w_{s_1} K(x, t_{s_1}) \sum_{k=0}^{M} y_k F_k(t_{s_1} - \tau), \qquad (8.5)$$

where t_{s_1}, $s_1 = 0, 1, ..., M$ are Legendre-Gauss points, i.e., the roots of $L_{M+1}(.) = 0$ and w_{s_1} are the corresponding weights defined as

$$w_{s_1} = \frac{2}{(1 - t_{s_1}^2)[L'_{M+1}(t_{s_1})]^2}, \qquad 0 \leq s_1 \leq M.$$

Substituting eq. (8.5) in eq. (8.4), we have

$$\begin{aligned}\sum_{i=0}^{m} p_i(x) \sum_{k=0}^{M} y_k F_k^{(i)}(x) + \sum_{j=0}^{n} p_j^*(x) \sum_{k=0}^{M} y_k F_k^{(j)}(x-\tau) \\ = g(x) + \sum_{s_1=0}^{M} w_{s_1} K(x, t_{s_1}) \sum_{k=0}^{M} y_k F_k(t_{s_1} - \tau). \qquad (8.6)\end{aligned}$$

Now, utilizing the eq. (8.6) at $M+1$ collocation points of Legendre-Gauss points $\{x_s\}_{s=0}^{M}$, we have

$$\begin{aligned}\sum_{i=0}^{m} p_i(x_s) \sum_{k=0}^{M} y_k F_k^{(i)}(x_s) + \sum_{j=0}^{n} p_j^*(x_s) \sum_{k=0}^{M} y_k F_k^{(j)}(x_s-\tau) \\ = g(x_s) + \sum_{s_1=0}^{M} w_{s_1} K(x_s, t_{s_1}) \sum_{k=0}^{M} y_k F_k(t_{s_1} - \tau). \qquad (8.7)\end{aligned}$$

Again we consider the mixed condition defined in eq. (8.2) and approximate by Lagrange interpolation formula (8.3) yielding

$$\sum_{i=0}^{m-1}\left[\alpha_{il}\sum_{k=0}^{M} y_k F_k^{(i)}(a)+\beta_{il}\sum_{k=0}^{M} y_k F_k^{(i)}(b)+\gamma_{il}\sum_{k=0}^{M} y_k F_k^{(i)}(c)\right]=\mu_i \tag{8.8}$$

for $l = 0, 1, ..., m-1$.
Combining eqs. (8.7) and (8.8), we get a system of algebraic equations with y_k, $k = 0, 1, ..., M$ unknowns. Numerically solving this system by Newton's method, we can determine the unknowns y_k, $k = 0, 1, ..., M$ and hence obtain the approximate solution $y(x) \approx \sum_{k=0}^{M} y_k F_k(x)$.

8.2.3 Illustrative examples

Example 8.2.1. *Let us consider the integro-differential-difference equation with variable coefficients*

$$y'(x) - y(x) + xy'(x-1) + y(x-1) = (x-2) + \int_{-1}^{1}(x+t)y(t-1)dt$$

with mixed condition

$$y(-1) - 2y(0) + y(1) = 0.$$

Solving the above equation by Legendre spectral collocation method for $M = 2$, we have obtained the unknowns as

$$y_0 = 1.67621, \quad y_1 = 4.00000, \quad y_2 = 6.32379.$$

Hence the approximate solution is

$$\begin{aligned} y(x) = \sum_{k=0}^{2} y_k F_k(x) &= y_0F_0(x) + y_1F_1(x) + y_2F_2(x)\\ &= 4 + 3x - 8.88178 \times 10^{-16}x^2\\ &\approx 4 + 3x \end{aligned}$$

The exact solution of this problem is $3x + 4$.

Example 8.2.2. *Let us consider the integro-differential difference equation with variable coefficients as [166, 103]*

$$\begin{aligned} y'''(x) - xy'(x) + y''(x-1) - xy(x-1) &= -(x+1)(\sin(x-1) + \cos x)\\ &\quad - \cos 2 + 1 + \int_{-1}^{1} y(t-1)dt, \end{aligned}$$

with conditions

$$y(0) = 0, \quad y'(0) = 1, \quad y''(0) = 0.$$

The exact solution of this problem is $y(x) = \sin x$. The obtained results by *the presented method have been compared with the results obtained by Legendre polynomials [166] and the Fibonacci collocation method (FCM) [103], and these numerical results along with the exact results are shown in Table 8.1. Table 8.2 cites the L_∞ error.*

TABLE 8.1: Numerical results along with exact results for Example 8.2.2

x	Exact	LSCM		FCM [103]	
		$M=6$	$M=7$	$N=8$	$N=9$
-1	-0.841471	-0.866814	-0.83644	-1.114125	-3.078521
-0.8	-0.717356	-0.729305	-0.71498	-0.869866	-1.875847
-0.6	-0.564642	-0.569211	-0.563732	-0.633677	-1.054038
-0.4	-0.389418	-0.390626	-0.389177	-0.4110374	-0.5333391
-0.2	-0.198669	-0.198802	-0.198643	-0.2014897	-0.2163871
0	0	0	0	0	0
0.2	0.0.198669	0.198769	0.19865	0.2016525	0.2155338
0.4	0.389418	0.390107	0.389294	0.4137037	0.5200945
0.6	0.564642	0.566704	0.564314	0.6476974	0.9903835
0.8	0.717356	0.721914	0.716785	0.9164897	1.689322
1	0.841471	0.850444	0.840739	1.235210	2.667579

TABLE 8.2: L_∞ error for example 8.2.2

Error	LSCM		FCM [103]		Legendre polynomials [166]	
	$M=6$	$M=7$	$N=8$	$N=9$	$m=6$	$m=7$
L_∞	2.53434E-2	5.03053E-3	3.937393E-1	2.23705E-0	3.84E-2	5.05E-3

Example 8.2.3. *Let us consider the integro-differential difference equation with variable coefficients as [213]*

$$
\begin{aligned}
(x+4)^2y''(x) &- (x+4)y'(x) + y(x-1) - y'(x-1) \\
&= \ln(x+3) - \frac{1}{x+3} + 3\ln(3) - 5\ln(5) + \int_{-1}^{1} y(t)dt,
\end{aligned}
$$

with conditions

$$
y(0) = \ln(4), \quad y'(0) = \frac{1}{4}.
$$

The exact solution of this problem is $y(x) = \ln(x+4)$. The results obtained by the presented method have been compared with the results obtained by the Boubaker polynomial [213] and these numerical results along with the exact results are shown in Table 8.3. Table 8.4 cites the L_∞ error.

Example 8.2.4. *Let us consider the integro-differential difference equation with variable coefficients as [76]*

$$
\begin{aligned}
y''(x) - xy'(x) + xy(x) - y'(x-1) + y(x-1) &= x(\sin x + \cos x) - \cos x \\
&+ \sin(x-1) + \cos(x-1) + 4x\sin(1) \\
&+ \int_{-1}^{1} (3t - 2x)y(t)dt,
\end{aligned}
$$

TABLE 8.3: Numerical results along with exact results for Example 8.2.3

x	Exact	LSCM		Boubaker polynomial [213]	
		$M=6$	$M=7$	$N=6$	$N=7$
-1	1.09861	1.09861	1.09861	1.098596	1.098657
-0.9	1.1314	1.1314	1.1314	1.131387	1.131431
-0.8	1.16315	1.16315	1.16315	1.163138	1.163173
-0.7	1.19392	1.19392	1.19392	1.193911	1.193937
-0.6	1.22378	1.22378	1.22378	1.223766	1.223787
-0.5	1.25276	1.25276	1.25276	1.252755	1.252771
-0.4	1.28093	1.28093	1.28093	1.280928	1.280939
-0.3	1.30833	1.30833	1.30833	1.308329	1.308336
-0.2	1.335	1.335	1.335	1.335	1.335001
-0.1	1.36098	1.36098	1.36098	1.360976	1.360977
0	1.38629	1.38629	1.38629	1.386294	1.386294

TABLE 8.4: L_∞ error for Example 8.2.3

Error	LSCM		Boubaker polynomial [213]	
	$M=6$	$M=7$	$N=6$	$N=7$
L_∞	5.86528E-7	1.77093E-6	1.66E-5	4.50E-5

with conditions

$$y(0)=1, \quad y'(0)=0.$$

The exact solution of this problem is $y(x)=\cos x$. *The results obtained by the presented method have been compared with the results obtained by the Leguerre collocation method (LCM) [76] and these numerical results along with the exact results are shown in Table 8.5. Table 8.6 cites the* L_∞ *error.*

Example 8.2.5. *Let us consider the integro-differential difference equation with variable coefficients as [92]*

$$y''(x)+xy'(x)+xy(x)+y'(x-1)+y(x-1)=e^{-x}+e+\int_{-1}^{0} ty(t-1)dt,$$

with conditions

$$y(0)=1, \quad y'(0)=-1.$$

The exact solution of this problem is $y(x)=e^{-x}$. *The results obtained by the presented method have been compared with the results obtained by the homotopy analysis method (HAM) [92] and these numerical results along with the exact results are shown in Table 8.7. Table 8.8 cites the absolute errors.*

TABLE 8.5: Numerical results along with exact results for Example 8.2.4

x	Exact	LSCM		LCM [76]	
		$M = 6$	$M = 7$	$N = 12$	$N = 12, M = 15$
0	1	1	1	0.999999	1
0.1	0.995004	0.995004	0.995004	0.996593	0.996600
0.2	0.980067	0.980066	0.980066	0.986641	0.986365
0.3	0.955336	0.955333	0.955334	0.969544	0.969444
0.4	0.921061	0.921052	0.921057	0.946139	0.946036
0.5	0.877583	0.877566	0.877577	0.916382	0.916381
0.6	0.825336	0.825314	0.825328	0.880504	0.880768
0.7	0.764842	0.764827	0.76483	0.838776	0.839522
0.8	0.696707	0.696723	0.696685	0.791509	0.793010
0.9	0.62161	0.621705	0.62161	0.739046	0.741629
1	0.540302	0.540552	0.540205	0.681764	0.685810

TABLE 8.6: L_∞ error for Example 8.2.4

Error	LSCM		LCM [76]	
	$M = 6$	$M = 7$	$N = 12$	$N = 12, M = 15$
L_∞	2.50083E-4	9.68491E-5	1.4146E-2	2.023575E-3

8.3 Legendre spectral collocation method for the solution of the model describing biological species living together

Mathematical modeling of real-life problems usually results in functional equations, e.g. partial differential equations, integral and integro-differential equations, stochastic equations and others. Many mathematical formulations of physical phenomena contain integro-differential equations; these equations arise in fluid dynamics, biological models and chemical kinetics. Integro-differential equations arise in many physical processes, such as the glass-forming process [203], nano-hydrodynamics [212], drop wise condensation [198], wind ripple in the desert [45] and biological modeling [193].

In this section, we consider the following system of integro-differential equations as

$$\begin{aligned}\frac{dx(t)}{dt} =& x(t)\left[k_1 - \gamma_1 y(t) - \int_{t-T_0}^{t} f_1(t-\tau)y(\tau)d\tau\right] + g_1(t),\\ & 0 \le t \le l, \quad k_1, \gamma_1 > 0,\end{aligned} \tag{8.9}$$

TABLE 8.7: Numerical results along with exact results for Example 8.2.5

x	Exact	LSCM		HAM [92]	
		$M = 10$	$M = 12$	$m = 10$	$m = 15$
-1	2.71828	2.71874	2.71832	2.71636	2.71821
-0.8	2.22554	2.22572	2.22556	1.82165	1.82210
-0.6	1.82212	1.82216	1.82212	1.82165	1.82210
-0.4	1.49182	1.49182	1.49182	1.49170	1.49182
-0.2	1.2214	1.2214	1.2214	1.22140	1.22140
0	1	1	1	1	1

TABLE 8.8: Absolute errors for Example 8.2.5

x	LSCM		HAM [92]	
	$M = 10$	$M = 12$	$m = 10$	$m = 15$
-1	4.58316E-4	3.88267E-5	1.92193E-3	7.42184E-5
-0.8	1.80503E-4	1.52916E-5	1.06345E-3	4.07102E-5
-0.6	4.3815E-5	3.71185E-6	4.71496E-4	1.54921E-5
-0.4	3.97513E-6	3.36791E-7	1.2820E-4	4.40566E-7
-0.2	6.31221E-6	5.34763E-7	3.25198E-4	2.57853E-6
0	0	0	0	0

$$\frac{dy(t)}{dt} = y(t)\left[-k_2 + \gamma_2 x(t) + \int_{t-T_0}^{t} f_2(t-\tau)x(\tau)d\tau\right] + g_2(t),$$
$$0 \le t \le l, \quad k_2, \gamma_2 > 0, \tag{8.10}$$

with initial conditions

$$x(0) = \alpha_1, \quad y(0) = \alpha_2,$$

where g_1, g_2, f_1, f_2 are given functions and $x(t)$, $y(t)$ are unknown functions.

Here $x(t)$ and $y(t)$ are the number of two separate species at time t, where the first species increases and the second decreases. If they are put together, assuming that the second species will feed on the first, there will be an increase in the rate of the second species $\frac{dy}{dt}$ which depends not only on the present populations $x(t)$, but also on all previous values of the first species. When a steady state condition is replaced between these two species, it is described by the following system of two integro-differential equations

$$\frac{dx(t)}{dt} = x(t)\left[k_1 - \gamma_1 y(t) - \int_{t-T_0}^{t} f_1(t-\tau)y(\tau)d\tau\right], \quad k_1 > 0, \tag{8.11}$$

$$\frac{dy(t)}{dt} = y(t)\left[-k_2 + \gamma_2 x(t) + \int_{t-T_0}^{t} f_2(t-\tau)x(\tau)d\tau\right], \quad k_2 > 0, \tag{8.12}$$

where k_1 and $-k_2$ are the coefficients of increasing and decreasing of the first and second species, respectively. The parameters γ_1, f_1 and γ_2, f_2 depend on the respective species. T_0 is assumed to be the finite heredity duration of both species. The system of integro-differential eqs. (8.11)-(8.12) is a special case of eqs. (8.9)-(8.10) with $g_1(t) = g_2(t) = 0$. The detailed formulations of eqs. (8.11)-(8.12) can be found in [89].

In this chapter, the aforesaid biological model has been solved by the Legendre spectral collocation method (LSCM) and the Bernstein polynomial collocation method (BPCM). The Legendre spectral collocation method has been applied by many authors to solve integral equations and integro-differential equations [200, 187, 51]. In this method, the unknown functions have been approximated by Lagrange polynomial with the Gauss-Legendre points as node points on the interval $[-1, 1]$. In general, the integro-differential equation defined over interval $[a, b]$ can be transformed into the interval $[-1, 1]$. The Bernstein polynomial collocation method has been applied to solve many kinds of integral equations and integro-differential equations [183, 128, 31, 150, 34]. The present methods reduce the integro-differential equations to a system of algebraic equations and then this system has been solved numerically. Also, the present method solutions are compared with other methods solutions in this present chapter. Illustrative examples have been discussed to demonstrate the validity and applicability of the proposed technique.

8.3.1 Numerical scheme by Legendre spectral collocation method

In this section, the basic idea of the Legendre spectral collocation method for solving integro-differential equations system defined in eqs. (8.9)-(8.12) has been presented. This idea has been proposed in [200, 187, 51]. In this procedure of approximation, the Legendre-Guass quadrature rule together with Lagrange interpolation polynomials have been used.

In order to use the Legendre spectral collocation method, we consider the Legendre-Gauss points $\{t_j\}_{j=0}^{M}$ i.e., the roots of $L_{M+1}(t) = 0$ where L_{M+1} is the $(M+1)^{th}$ Legendre polynomial.

Let us consider the system of integro-differential equations defined in eqs. (8.9)-(8.10) and approximate the unknown functions $x(t)$ and $y(t)$ by using eq. (8.3) as

$$x(t) = \sum_{k=0}^{M} x_k F_k(t), \tag{8.13}$$

$$y(t) = \sum_{k=0}^{M} y_k F_k(t). \tag{8.14}$$

Now, eqs. (8.9)-(8.10) can be reduced as

$$\sum_{k=0}^{M} x_k F_k'(t) = \sum_{k=0}^{M} x_k F_k(t) \left[k_1 - \gamma_1 \sum_{k=0}^{M} y_k F_k(t) - \int_{t-T_0}^{t} f_1(t-\tau) \left(\sum_{k=0}^{M} y_k F_k(\tau) \right) d\tau \right] + g_1(t), \quad (8.15)$$

$$\sum_{k=0}^{M} y_k F_k'(t) = \sum_{k=0}^{M} y_k F_k(t) \left[-k_2 + \gamma_2 \sum_{k=0}^{M} x_k F_k(t) + \int_{t-T_0}^{t} f_2(t-\tau) \left(\sum_{k=0}^{M} x_k F_k(\tau) \right) d\tau \right] + g_2(t). \quad (8.16)$$

Now, we consider the integral part of eq. (8.15)-(8.16), and in order to use the Gauss-Legendre quadrature rule, we have to change the interval $[t - T_0, t]$ to $[-1, 1]$ by the transformation

$$s = 1 + 2\left(\frac{\tau - t}{T_0}\right).$$

Now, the integrands of eqs. (8.15)-(8.16) can be reduced as

$$\begin{aligned} &\int_{t-T_0}^{t} f_1(t-\tau) \left(\sum_{k=0}^{M} y_k F_k(\tau) \right) d\tau \\ &= \frac{T_0}{2} \int_{-1}^{1} f_1\left(-\frac{T_0}{2}(s-1)\right) \left(\sum_{k=0}^{M} y_k F_k\left(t + \frac{T_0}{2}(s-1)\right)\right) ds \\ &= \frac{T_0}{2} \sum_{j=0}^{M} w_j f_1\left(-\frac{T_0}{2}(s_j-1)\right) \left(\sum_{k=0}^{M} y_k F_k\left(t + \frac{T_0}{2}(s_j-1)\right)\right) \quad (8.17) \end{aligned}$$

$$\begin{aligned} &\int_{t-T_0}^{t} f_2(t-\tau) \left(\sum_{k=0}^{M} x_k F_k(\tau) \right) d\tau \\ &= \frac{T_0}{2} \int_{-1}^{1} f_2\left(-\frac{T_0}{2}(s-1)\right) \left(\sum_{k=0}^{M} x_k F_k\left(t + \frac{T_0}{2}(s-1)\right)\right) ds \\ &= \frac{T_0}{2} \sum_{j=0}^{M} w_j f_2\left(-\frac{T_0}{2}(s_j-1)\right) \left(\sum_{k=0}^{M} x_k F_k\left(t + \frac{T_0}{2}(s_j-1)\right)\right) \quad (8.18) \end{aligned}$$

where s_j, $j = 0, 1, ..., M$ are Legendre-Gauss points; i.e., the roots of $L_{M+1}(t) = 0$ and w_j are the corresponding weights defined as

$$w_j = \frac{2}{(1 - s_j^2)[L'_{M+1}(s_j)]^2}, \quad 0 \le j \le M.$$

Using eqs. (8.17)-(8.18) in eqs. (8.15)-(8.16), we have

$$\sum_{k=0}^{M} x_k F_k'(t) = \sum_{k=0}^{M} x_k F_k(t) \left[k_1 - \gamma_1 \sum_{k=0}^{M} y_k F_k(t) \right.$$
$$\left. -\frac{T_0}{2} \sum_{j=0}^{M} w_j f_1 \left(-\frac{T_0}{2}(s_j - 1) \right) \left(\sum_{k=0}^{M} y_k F_k \left(t + \frac{T_0}{2}(s_j - 1) \right) \right) \right] + g_1(t) \tag{8.19}$$

$$\sum_{k=0}^{M} y_k F_k'(t) = \sum_{k=0}^{M} y_k F_k(t) \left[-k_2 + \gamma_2 \sum_{k=0}^{M} x_k F_k(t) \right.$$
$$\left. +\frac{T_0}{2} \sum_{j=0}^{M} w_j f_2 \left(-\frac{T_0}{2}(s_j - 1) \right) \left(\sum_{k=0}^{M} x_k F_k \left(t + \frac{T_0}{2}(s_j - 1) \right) \right) \right] + g_2(t) \tag{8.20}$$

Now, applying the collocation points as Gauss-Legendre points t_i, $i = 0, 1, ..., M-1$ to the eqs. (8.19)-(8.20), we obtain

$$\sum_{k=0}^{M} x_k F_k'(t_i) = \sum_{k=0}^{M} x_k F_k(t_i) \left[k_1 - \gamma_1 \sum_{k=0}^{M} y_k F_k(t_i) \right.$$
$$\left. -\frac{T_0}{2} \sum_{j=0}^{M} w_j f_1 \left(-\frac{T_0}{2}(s_j - 1) \right) \left(\sum_{k=0}^{M} y_k F_k \left(t_i + \frac{T_0}{2}(s_j - 1) \right) \right) \right] + g_1(t_i) \tag{8.21}$$

$$\sum_{k=0}^{M} y_k F_k'(t_i) = \sum_{k=0}^{M} y_k F_k(t_i) \left[-k_2 + \gamma_2 \sum_{k=0}^{M} x_k F_k(t_i) \right.$$
$$\left. +\frac{T_0}{2} \sum_{j=0}^{M} w_j f_2 \left(-\frac{T_0}{2}(s_j - 1) \right) \left(\sum_{k=0}^{M} x_k F_k \left(t_i + \frac{T_0}{2}(s_j - 1) \right) \right) \right] + g_2(t_i) \tag{8.22}$$

Equations (8.21)-(8.22) constitute a system of $2M$ number of nonlinear algebraic equations with $2M + 2$ number of unknowns. Again from the initial conditions, we have

$$\sum_{k=0}^{M} x_k F_k(0) = \alpha_1, \tag{8.23}$$

$$\sum_{k=0}^{M} y_k F_k(0) = \alpha_2. \tag{8.24}$$

Thus eqs. (8.21)-(8.24) form a system of $2M+2$ number of nonlinear algebraic equations with $2M+2$ number of unknowns for x_k and y_k for $k = 0, 1, ..., M$. Solving this system numerically, we can obtain the values for unknowns x_k and y_k $k = 0, 1, ..., M$. Hence we obtain the approximate solutions $x(t)$ and $y(t)$ using the eqs. (8.13)-(8.14). The obtained results are then compared with the results obtained by the Bernstein polynomial collocation method and other methods available in literature. The methodology using Bernstein polynomials is given below.

- **Numerical scheme by Bernstein polynomials collocation method**

Bernstein polynomials defined in Chapter 2 form a complete basis [183, 36] over the interval $[a, b]$. For solving eqs. (8.9)-(8.10), we first approximate the unknown functions $x(t)$ and $y(t)$ as

$$x(t) = C^T B(t), \qquad y(t) = D^T B(t) \tag{8.25}$$

where C and D are defined the same as in Section 2.7.2. The eqs. (8.9)-(8.10) can be reduced as

$$C^T B'(t) = (C^T B(t)) \left[k_1 - \gamma_1(D^T B(t)) - \int_{t-T_0}^{t} f_1(t-\tau)(D^T B(\tau))d\tau\right] + g_1(t) \tag{8.26}$$

$$D^T B'(t) = (D^T B(t)) \left[-k_2 + \gamma_2(C^T B(t)) + \int_{t-T_0}^{t} f_2(t-\tau)(C^T B(\tau))d\tau\right] + g_2(t) \tag{8.27}$$

Now, substituting the collocation points $t_l = t_0 + lh$, where $t_0 = a$, $h = \frac{b-a}{n}$ and $l = 0, 1, ..., n-1$, in eqs. (8.26)-(8.27), we have

$$\begin{aligned} C^T B'(t_l) = (C^T B(t_l)) \Big[k_1 - \gamma_1(D^T B(t_l)) \\ - \int_{t_l-T_0}^{t_l} f_1(t_l-\tau)(D^T B(\tau))d\tau\Big] + g_1(t_l) \end{aligned} \tag{8.28}$$

$$\begin{aligned} D^T B'(t_l) = (D^T B(t_l)) \Big[-k_2 + \gamma_2(C^T B(t_l)) \\ + \int_{t_l-T_0}^{t_l} f_2(t_l-\tau)(C^T B(\tau))d\tau\Big] + g_2(t_l) \end{aligned} \tag{8.29}$$

Again from boundary conditions, we get

$$\left.\begin{aligned} C^T B(0) = \alpha_1 \\ D^T B(0) = \alpha_2 \end{aligned}\right\} \tag{8.30}$$

Thus, from eqs. (8.28)-(8.30), we obtain a system of $2n+2$ number of nonlinear algebraic equations with the same number of unknowns as C and D. Solving this system numerically, we obtain the values of C and D and hence obtain the approximate solutions using eq. (8.25).

TABLE 8.9: Comparison of numerical results for Example 8.3.1

t	Absolute error for $x(t)$			Absolute error for $y(t)$		
	LPCM	BPCM	VIM[193]	LPCM	BPCM	VIM[193]
0.1	0	1.41E-11	3.15E-4	1.38E-17	1.22E-12	3.34E-5
0.2	3.33E-16	2.15E-11	4.27E-4	2.77E-17	2.96E-12	8.54E-5
0.3	7.21E-16	2.58E-11	4.72E-4	0	4.16E-12	1.33E-4
0.4	1.33E-15	2.85E-11	4.85E-4	1.38E-16	5.12E-12	1.79E-4
0.5	1.99E-15	2.98E-11	4.74E-4	4.44E-16	6.58E-12	2.22E-4
0.6	2.99E-15	2.84E-11	4.45E-4	8.88E-16	9.19E-12	2.37E-4
0.7	4.21E-15	2.29E-11	4.36E-4	1.58E-15	1.29E-11	1.62E-4
0.8	5.77E-15	1.22E-11	4.35E-4	2.38E-15	1.69E-11	1.07E-4
0.9	6.88E-15	3.33E-12	9.10E-4	3.27E-15	1.92E-11	7.31E-4
1.0	7.54E-15	2.12E-11	1.82E-3	3.74E-15	1.69E-11	1.90E-3

8.3.2 Illustrative examples

Example 8.3.1. *Consider the system of integro-differential equations [193] defined in eqs.* (8.9)-(8.10) *with*

$$
\begin{aligned}
&f_1(t) = 1, \quad f_2(t) = t - 1,\\
&k_1 = 1, \quad k_2 = 2,\\
&\gamma_1 = \frac{1}{3}, \quad \gamma_2 = 1,\\
&T_0 = \frac{1}{2},\\
&\alpha_1 = 1, \quad \alpha_2 = 0,\\
&g_1(t) = -\frac{5}{2}t^3 + \frac{49}{12}t^2 + \frac{17}{12}t - \frac{23}{6},
\end{aligned}
$$

and

$$g_2(t) = \frac{15}{8}t^3 - \frac{1}{4}t^2 + \frac{3}{8}t - 1.$$

The exact solution of this problem is $x(t) = -3t + 1$ *and* $y(t) = t^2 - t$. *The numerical results obtained by LSCM for* $M = 6$ *have been compared with the results obtained by BPCM* ($n = 6$) *and VIM [193]. The comparisons have been cited in Table 8.9.*

Example 8.3.2. *Consider the system of integro-differential equations [193]*

TABLE 8.10: Comparison of numerical results for Example 8.3.2

t	Absolute error for $x(t)$			Absolute error for $y(t)$		
	LPCM	BPCM	VIM[193]	LPCM	BPCM	VIM[193]
0.1	7.73E-12	5.27E-8	4.50E-10	6.23E-11	1.54E-6	9.80E-8
0.2	2.35E-11	8.03E-8	4.07E-9	1.72E-10	1.36E-6	6.93E-8
0.3	3.13E-11	1.00E-7	4.72E-8	1.67E-10	1.05E-6	2.69E-7
0.4	2.52E-11	1.20E-7	3.64E-7	8.52E-12	8.60E-7	3.55E-7
0.5	2.04E-11	1.41E-7	2.03E-6	1.62E-10	7.06E-7	2.49E-6
0.6	3.86E-11	1.66E-7	8.80E-6	6.70E-12	5.50E-7	1.08E-5
0.7	6.76E-11	1.95E-7	3.12E-5	3.26E-10	4.30E-7	3.85E-5
0.8	4.87E-11	2.29E-7	9.44E-5	4.45E-14	3.57E-7	1.14E-4
0.9	3.22E-11	2.72E-7	2.51E-4	6.05E-10	2.57E-7	3.00E-4
1.0	7.77E-10	3.21E-7	6.04E-4	8.60E-9	1.31E-7	7.11E-4

defined in eqs. (8.9)-(8.10) *with*

$$f_1(t) = 2t - 3, \quad f_2(t) = t,$$
$$k_1 = 2, \quad k_2 = 2,$$
$$\gamma_1 = 1, \quad \gamma_2 = 1,$$
$$T_0 = \frac{1}{3},$$
$$\alpha_1 = 0, \quad \alpha_2 = 0,$$
$$g_1(t) = t^2\left(2 - 3te^{-t} - \frac{7}{2}e^{-t} + \frac{13}{6}te^{\frac{1}{3}-t} + \frac{22}{9}e^{\frac{1}{3}-t}\right) - 2t,$$

and

$$g_2(t) = \frac{1}{648}e^{-t}\left(342t^3 - 8t^2 + 325t + 324\right).$$

The exact solution of this problem is $x(t) = -t^2$ *and* $y(t) = \frac{1}{2}te^{-t}$. *The numerical results obtained by LSCM for* $M = 10$ *have been compared with the results obtained by BPCM* $(n = 6)$ *and VIM [193]. The comparisons have been cited in Table 8.10.*

Example 8.3.3. *Consider the system of integro differential equations [193]*

TABLE 8.11: Comparison of numerical results for Example 8.3.3

t	Absolute error for $x(t)$			Absolute error for $y(t)$		
	LPCM	BPCM	VIM[193]	LPCM	BPCM	VIM[193]
0.1	2.73E-12	2.40E-7	5.22E-10	2.63E-12	2.11E-7	4.63E-10
0.2	8.37E-12	2.69E-7	6.23E-9	7.78E-12	2.18E-7	2.88E-9
0.3	1.02E-11	2.76E-7	1.59E-7	8.81E-12	2.02E-7	7.52E-8
0.4	4.76E-12	2.95E-7	1.24E-6	2.46E-12	1.98E-7	6.09E-7
0.5	1.89E-12	3.20E-7	6.12E-6	4.36E-12	1.97E-7	2.97E-6
0.6	2.75E-12	3.45E-7	2.20E-5	7.45E-13	1.95E-7	1.06E-5
0.7	1.72E-11	3.75E-7	6.39E-5	1.49E-11	1.94E-7	3.06E-5
0.8	1.00E-11	4.14E-7	1.58E-4	5.31E-12	1.99E-7	7.54E-5
0.9	1.52E-11	4.54E-7	3.48E-4	2.09E-11	2.02E-7	1.64E-4
1.0	3.05E-10	4.93E-7	6.94E-4	3.25E-10	2.01E-7	3.24E-4

defined in eqs. (8.9)-(8.10) *with*

$$f_1(t) = 1, \quad f_2(t) = e^{-t},$$
$$k_1 = \frac{1}{3}, \quad k_2 = \frac{1}{2},$$
$$\gamma_1 = 2, \quad \gamma_2 = 1,$$
$$T_0 = \frac{3}{10},$$
$$\alpha_1 = 0, \quad \alpha_2 = 0,$$
$$g_1(t) = \frac{1}{4}\cos t - \frac{1}{4}\sin t\left(\frac{1}{3} + \frac{1}{2}\sin t - \frac{1}{4}\cos t + \frac{1}{4}\cos(t-3/10)\right),$$

and

$$g_2(t) = -\frac{1}{4}\cos t + \frac{1}{4}\sin t\left(-\frac{1}{2} + \frac{3}{8}\sin t - \frac{1}{8}\cos t + \frac{1}{8}e^{-3/10}(\cos(t-3/10) - \sin(t-3/10))\right).$$

The exact solution of this problem is $x(t) = \frac{1}{4}\sin t$ *and* $y(t) = -\frac{1}{4}\sin t$. *The numerical results obtained by LSCM for* $M = 10$ *have been compared with the results obtained by BPCM* ($n = 6$) *and VIM [193]. The comparisons have been cited in Table 8.11.*

Example 8.3.4. *Consider the system of integro-differential equations [193]*

TABLE 8.12: Comparison of numerical results for Example 8.3.4

t	Absolute error for $x(t)$			Absolute error for $y(t)$		
	LPCM	BPCM	VIM[193]	LPCM	BPCM	VIM[193]
0.1	3.58E-8	1.24E-6	3.59E-7	2.34E-6	4.51E-4	1.09E-5
0.2	1.27E-7	4.36E-6	2.66E-7	6.60E-6	4.22E-4	1.55E-5
0.3	2.28E-7	8.83E-6	4.66E-7	6.66E-6	3.43E-4	8.22E-6
0.4	2.88E-7	1.40E-5	1.64E-5	2.23E-7	2.89E-4	9.26E-5
0.5	2.82E-7	1.95E-5	6.93E-5	5.67E-6	2.44E-4	3.95E-4
0.6	2.47E-7	2.50E-5	1.73E-4	2.30E-7	1.99E-4	9.37E-4
0.7	2.83E-7	3.02E-5	3.23E-4	1.22E-5	1.65E-4	1.62E-3
0.8	4.70E-7	3.52E-5	4.92E-4	1.47E-6	1.44E-4	2.26E-3
0.9	5.88E-7	4.01E-5	6.41E-4	2.14E-5	1.20E-4	2.63E-3
1.0	5.10E-7	4.48E-5	7.37E-4	2.89E-4	9.96E-4	2.57E-3

defined in eqs. (8.9)-(8.10) *with*

$$f_1(t) = t, \quad f_2(t) = t + 1,$$
$$k_1 = 1, \quad k_2 = 1,$$
$$\gamma_1 = \frac{1}{2}, \quad \gamma_2 = 3,$$
$$T_0 = \frac{1}{4},$$
$$\alpha_1 = 0, \quad \alpha_2 = -1,$$
$$g_1(t) = 2t - 1 - (t^2 - t)\left(1 + \frac{11}{18}e^{-3t} - \frac{1}{36}e^{\frac{3}{4}-3t}\right),$$

and

$$g_2(t) = \frac{1}{3072}e^{-3t}\left(10080t^2 - 10304t + 6275\right).$$

The exact solution of this problem is $x(t) = t^2 - t$ *and* $y(t) = -e^{-3t}$. *The numerical results obtained by LSCM for* $M = 10$ *have been compared with the results obtained by BPCM* ($n = 6$) *and VIM [193]. The comparisons have been cited in Table 8.12.*

Example 8.3.5. *The last example is an example of biological species living together [21, 37] defined in eqs.* (8.11)-(8.12).
Let consider the eqs. (8.11)-(8.12) *with* $k_1 = 0.02$, $k_2 = 0.01$, $\alpha_1 = \alpha_2 = 10$, $\gamma_1 = \gamma_2 = 0.01$, *and* $T_0 = 0.1$. *The results obtained by present methods (LSCM* ($M = 10$) *and BPCM* ($n = 6$)*) have been examined by the results obtained by ADM [21] and HPM [37] for some values of* t, *and the comparisons have been demonstrated in Table 8.13.*

From Tables 8.9-8.12, it can be easily observed that LPCM provides more accurate and better solutions than other methods. Table 8.13 cites that results

TABLE 8.13: Comparison of numerical results for Example 8.3.5

t	$x(t)$				$y(t)$			
	LPCM	BPCM	ADM [21]	HPM [37]	LPCM	BPCM	ADM [21]	HPM [37]
0.01	9.90181	9.90199	9.897	9.8968	10.1088	10.1086	10.116	10.1042
0.03	9.705	9.70547	9.691	9.6904	10.3268	10.3262	10.349	10.3125
0.05	9.50777	9.50846	9.484	9.4838	10.5451	10.5443	10.580	10.5208
0.07	9.3103	9.31113	9.278	9.2771	10.7636	10.7627	10.813	10.7291
0.1	9.01399	9.01497	8.968	8.9668	11.0913	11.0902	11.162	11.0416
0.3	7.07726	7.07823	6.905	6.8911	13.2279	13.2267	13.485	13.1250
0.5	5.31809	5.31893	4.842	4.8030	15.16	15.159	15.808	15.2082
0.7	3.84391	3.84465	2.780	2.7023	16.7701	16.7692	18.131	17.2916

obtained by LPCM and BPCM agree quite satisfactorily and these results are also very close to the other method results.

8.4 Conclusion

In this work, the Legendre spectral collocation method has been applied to solve the Fredholm integro-differential-difference equations with variable coefficients and mixed conditions, and a system of integro-differential equations modeling biological species living together. This method is a very straightforward technique to solve these equations. The Legendre spectral collocation method transforms the integral equations to a system of algebraic equations which can be solved by usual numerical methods. The application of the Legendre-Gauss points is the main advantage of this presented method. In Section 8.3, the LSCM solutions are compared with the solutions obtained by the Bernstein collocation method and other methods. From the tables, it is clear that the results obtained by LSCM are more accurate than those of by other methods. LSCM may provide more accurate results by increasing the value of M. Also the Legendre spectral collocation method takes fractions of seconds to accomplish the computation. However, one may face the following difficulties with the proposed method applied in the present chapter: the first one is that it yields a nonlinear system of equations which may fail to converge without suitable initial guess and the second is that the collocation points have to be chosen suitably for obtaining an accurate solution. The illustrative examples have been included to demonstrate the validity and applicability of the proposed techniques. These examples also exhibit the accuracy and efficiency of the present schemes.

Chapter 9

Numerical Solutions of Fuzzy Integral Equations

9.1 Introduction

The study of fuzzy integral equations and fuzzy differential equations is an emerging area of research for many authors. Originally, the concept of fuzzy sets was first introduced by Zadeh [225, 50]. The development of fuzzy integral equations was first invented by Kaleva [90] and Seikkala [191]. In recent years, many researchers have focused their interest on this field and published many articles which are available in the literature. Many analytical methods like the Adomian decomposition method [23], homotopy analysis method [144], and homotopy perturbation method [10] have been used to solve fuzzy integral equations. There are available many numerical techniques to solve fuzzy integral equations. The method of successive approximations

[41, 43], quadrature rule [186], the Nystrom method [1], Lagrange interpolation [12], Bernstein polynomials [61], Chebyshev interpolation [30], Legendre wavelet method [185], Sinc function [97], the residual minimization method [87], the fuzzy transforms method [60], and the Galerkin method [116] have been applied to solve fuzzy integral equations numerically. Recently, Sadatrasoul et al. [170] have solved nonlinear fuzzy integral equations by applying the iterative method. Many theories related to fuzzy fractional functional integral and differential equations have been included in [149] and convergence in measure theorem for nonlinear integral functionals has been provided in [94]. The existence of solutions to fuzzy differential equations with generalized Hukuhara derivative via contractive-like mapping principles has been presented by Villamizar-Roa et al. [202]. A classical solution of the fuzzy boundary value problem has been given in [65]. The Cauchy problem for complex fuzzy differential equations has been solved in [93]. Hybrid block-pulse functions and the Taylor series method [25] have been applied to solve nonlinear fuzzy Fredholm integral equations of the second kind and also linear Fredholm fuzzy integral equations of the second kind have been solved by artificial neural networks [62]. Also, there are available many works related to fuzzy integro-differential equations [105, 106, 8, 2, 145] in the literature. The learned researcher Abbasbandy et al. have solved fuzzy integro-differential equations by the homotopy analysis method [2]. Fuzzy Fredholm integro-differential equations have been solved by the Newton-cotes method [145]. The existence and uniqueness of the solutions of fuzzy integro-differential equations have been presented in [105, 8, 154]. In this chapter, we have solved Hammerstein fuzzy integral equations, fuzzy Hammerstein Volterra delay integral equations and fuzzy integro-differential equations. In Section 9.2, we discuss the preliminaries of fuzzy calculus. In Section 9.3, a nonlinear fuzzy Hammerstein integral equation has been solved by Bernstein polynomials and Legendre wavelets, and then compared with the homotopy analysis method. In Section 9.4, we have solved nonlinear fuzzy Hammerstein Volterra integral equations with constant delay by the Bernoulli wavelet method and then compared with the B-spline wavelet method. In Section 9.5, the fuzzy integro-differential equation has been solved by Legendre wavelet method and compared with the homotopy analysis method, and Section 9.6 describes the concluding remarks.

9.2 Preliminaries of fuzzy integral equations

In this section, the most basic notations used in fuzzy calculus are introduced. We start with defining a fuzzy number.

Definition 9.2.1. ([67]) *A fuzzy number u is represented by an ordered pair of functions $(\underline{u}(r), \overline{u}(r))$; $0 \leq r \leq 1$ which satisfying the following properties.*

1. *$\underline{u}(r)$ is a bounded monotonic increasing left continuous function.*

2. *$\overline{u}(r)$ is a bounded monotonic decreasing left continuous function.*

3. *$\underline{u}(r) \leq \overline{u}(r), \quad 0 \leq r \leq 1.$*

For arbitrary $u(r) = (\underline{u}(r), \overline{u}(r))$, $v(r) = (\underline{v}(r), \overline{v}(r))$ and $k > 0$, we define addition $(u + v)$ and scalar multiplication by k as

i. $(\underline{u+v})\,(r) = \underline{u}(r) + \underline{v}(r)$

ii. $(\overline{u+v})\,(r) = \overline{u}(r) + \overline{v}(r)$

iii. $(\underline{ku})\,(r) = k\underline{u}(r)$, $(\overline{ku})\,(r) = k\overline{u}(r)$

Definition 9.2.2. ([211]) *For arbitrary fuzzy numbers $u, v \in E$, we use the distance*

$$D(u,v) = \sup_{0\leq r\leq 1} \left[\max\left\{|\overline{u}(r) - \overline{v}(r)|, |\underline{u}(r) - \underline{v}(r)|\right\}\right]$$

and it is shown that (E, D) is a complete metric space.

Remark 9.2.3. ([211]) *If the fuzzy function $f(t)$ is continuous in the metric D, its definite integral exists. Also*

$$\left(\underline{\int_a^b f(t;r)dt}\right) = \int_a^b \underline{f}(t;r)dt,$$

$$\left(\overline{\int_a^b f(t;r)dt}\right) = \int_a^b \overline{f}(t;r)dt$$

Definition 9.2.4. *If $f : R \longrightarrow E$ is a fuzzy function (where E is a subset of a Banach space) and $t_0 \in R$, the derivative $f'(t_0)$ of f at a point t_0 is defined by*

$$f'(t_0) = \lim_{h\to 0^+} \frac{f(t_0 + h) - f(t_0)}{h}, \tag{9.1}$$

provided that this limit taken with respect to the metric D, exists and $h > 0$ is a sufficiently small parameter.

The elements $f(t_0 + h)$ and $f(t_0)$ in eq. (9.1) are in Banach space $B = \overline{C}[0,1] \times \overline{C}[0,1]$. Thus, if $f(t_0 + h) = (\underline{a}, \overline{a})$ and $f(t_0) = (\underline{b}, \overline{b})$, then $f(t_0 + h) - f(t_0) = (\underline{a} - \underline{b}, \overline{a} - \overline{b})$.

Clearly, $[f(t_0 + h) - f(t_0)]/h$ may not be a fuzzy number for all h. However, if it approaches $f'(t_0)$ (in B) and $f'(t_0)$ is also a fuzzy number (in E), this number is the fuzzy derivative of $f(t)$ at t_0. In this case, if $f = (\underline{f}, \overline{f})$, then $f'(t_0) = (\underline{f}'(t_0), \overline{f}'(t_0))$, where $\underline{f}'$, $\overline{f}'$ are the classic derivative of $\underline{f}$, $\overline{f}$, respectively and $t_0 \in R$.

9.3 Numerical solution of fuzzy Hammerstein integral equations

In this section, we consider the nonlinear fuzzy Fredholm-Hammerstein integral equation of the form

$$u(t) = g(t) + \int_0^1 H(t,s)F(u(s))ds, \quad H(t,s) \in C([0,1]\times[0,1]), \quad t \in [0,1], \tag{9.2}$$

where u, g and F are fuzzy functions, and $H(t,s)$ is positive in $[0,1]$. Let

$$\begin{aligned} u(t,r) &= \left(\underline{u}(t,r), \overline{u}(t,r)\right), \\ g(t,r) &= \left(\underline{g}(t,r), \overline{g}(t,r)\right), \\ F(u(t,r)) &= \left(\underline{F}(u(t,r)), \overline{F}(u(t,r))\right). \end{aligned}$$

Eq. (9.2), in the crisp sense, is converted into a system as

$$\underline{u}(t,r) = \underline{g}(t,r) + \int_0^1 H(t,s)\underline{F}\left(u(s,r)\right)ds, \tag{9.3}$$

$$\overline{u}(t,r) = \overline{g}(t,r) + \int_0^1 H(t,s)\overline{F}\left(u(s,r)\right)ds. \tag{9.4}$$

Equations (9.3) and (9.4) have been solved by the Bernstein polynomial collocation method and the Legendre wavelet method and again compared with the homotopy analysis method.

9.3.1 Numerical scheme by Bernstein polynomial collocation method

Consider the eq. (9.3) for solving by the Bernstein polynomial collocation method, first approximate the unknown function $\underline{u}(t,r)$ by using two dimensional Bernstein polynomials as

$$\underline{u}(t,r) \approx \sum_{i=0}^{n_1}\sum_{j=0}^{n_2} c_{i,j}B_{i,n_1}(t)B_{j,n_2}(r) \tag{9.5}$$

Eq. (9.3) can be reduced as

$$\sum_{i=0}^{n_1}\sum_{j=0}^{n_2} c_{i,j}B_{i,n_1}(t)B_{j,n_2}(r) = \underline{g}(t,r) + \int_0^1 H(t,s)\underline{F}\left(\sum_{i=0}^{n_1}\sum_{j=0}^{n_2} c_{i,j}B_{i,n_1}(s)B_{j,n_2}(r)\right)ds \tag{9.6}$$

Utilizing eq. (9.6) with the collocation points t_l and r_m defined as

$$t_l = t_0 + lh_1, \quad t_0 = 0, \quad h_1 = \frac{1}{n_1}, \quad l = 0, 1, ..., n_1,$$

$$r_m = r_0 + mh_2, \quad r_0 = 0, \quad h_2 = \frac{1}{n_2}, \quad m = 0, 1, ..., n_2,$$

eq. (9.6) reduces to a system of $(n_1 + 1)(n_2 + 1)$ number of nonlinear algebraic equations with the same number of unknowns as $c_{i,j}$, $i = 0, 1, ..., n_1$, $j = 0, 1, ..., n_2$. This algebraic system has been solved by Newton's method to obtain the unknowns $c_{i,j}$, $i = 0, 1, ..., n_1$, $j = 0, 1, ..., n_2$. Hence we get the solution $\underline{u}(t, r)$ from eq. (9.5) and the same algorithm can be applied to obtain the approximate solution of $\overline{u}(t, r)$.

9.3.2 Numerical scheme by Legendre wavelet method

Consider the eq. (9.3) for solving by the Legendre wavelet method, first approximate the unknown function $\underline{u}(t, r)$ by using the two dimensional Legendre wavelets (see eq. (2.34) of Chapter 2) as

$$\underline{u}(t, r) = C^T \Psi(t, r). \tag{9.7}$$

Now, eq. (9.3) can be reduced as

$$C^T \Psi(t, r) = \underline{g}(t, r) + \int_0^1 H(t, s) \underline{F} \left(C^T \Psi(s, r) \right) ds. \tag{9.8}$$

In order to use the Gauss-Legendre integration formula for eq. (9.8), we transfer the interval $[0, 1]$ to $[-1, 1]$ by means of the transformation $\tau = 2s - 1$. Therefore, eq. (9.8) can be written as

$$C^T \Psi(t, r) = \underline{g}(t, r) + \frac{1}{2} \int_{-1}^1 H\left(t, \frac{\tau + 1}{2}\right) \underline{F}\left(C^T \Psi\left(\frac{\tau + 1}{2}, r\right)\right) d\tau. \tag{9.9}$$

By using the Gauss-Legendre integration formula, we get

$$C^T \Psi(t, r) = \underline{g}(t, r) + \frac{1}{2} \sum_{j=1}^{M} w_j H\left(t, \frac{\tau_j + 1}{2}\right) \underline{F}\left(C^T \Psi\left(\frac{\tau_j + 1}{2}, r\right)\right), \tag{9.10}$$

where τ_j are m zeros of Legendre polynomials P_{m+1} and w_j are the corresponding weights. Now we collocate the eq. (9.10) at $t_p = \frac{2p-1}{2^{k_1} M_1}$, $p = 1, 2, ..., 2^{k_1 - 1} M_1$ and $r_q = \frac{2q-1}{2^{k_2} M_2}$, $q = 1, 2, ..., 2^{k_2 - 1} M_2$, we have

$$C^T \Psi(t_p, r_q) = \underline{g}(t_p, r_q) + \frac{1}{2} \sum_{j=1}^{M} w_j H\left(t_p, \frac{\tau_j + 1}{2}\right) \underline{F}\left(C^T \Psi\left(\frac{\tau_j + 1}{2}, r_q\right)\right). \tag{9.11}$$

Eq. (9.11) gives a system of $2^{k_1-1}M_1 \times 2^{k_2-1}M_2$ number of algebraic equations with the same number of unknowns for C^T. Again solving this system numerically by Newton's method, we can find the value for unknowns for C^T and hence obtain the approximate solution for $\underline{u}(t,r)$. The same algorithm can be applied to obtain the approximate solution of $\overline{u}(t,r)$.

Theorem 9.3.1. *The series solution* $y(x) \cong \sum_{n=1}^{2^{k-1}} \sum_{m=0}^{M-1} c_{n,m}\psi_{n,m}(x)$ *defined in eq.* (2.31) *using Legendre wavelet method converges to* $y(x)$.

Proof. See Theorem 6.2.1 of Chapter 6. □

9.3.3 Illustrative examples

Example 9.3.1. *Consider the following nonlinear Hammerstein fuzzy integral equation*

$$u(t) = g(t) + \int_0^1 H(t,s)e^{-s}(u(s))^2 ds, \quad t \in [0,1] \tag{9.12}$$

with

$$H(t,s) = \begin{cases} \frac{1}{6}s^2(1-t)^2(3t-s-2ts), & 0 \le s \le t \le 1, \\ \frac{1}{6}t^2(1-s)^2(3s-t-2ts), & 0 \le t \le s \le 1, \end{cases}$$

and

$$g(t,r) = g_1(t,r) + g_2(t,r), \quad t, r \in [0,1],$$

where

$$g_1(t,r) = \left[(1-t)^2(3t+1)\left(1 - \frac{1-r}{10}\right), (1-t)^2(3t+1)\left(1 + \frac{1-r}{10}\right)\right],$$

$$g_2(t,r) = \left[t^2(2-t)\left(e - \frac{1-r}{10}\right), t^2(2-t)\left(e + \frac{1-r}{10}\right)\right].$$

Eq. (9.12), *in the crisp sense, converted into a system as*

$$\underline{u}(t,r) = \underline{g}(t,r) + \int_0^1 H(t,s)e^{-s}\left(\underline{u}(s,r)\right)^2 ds, \tag{9.13}$$

$$\overline{u}(t,r) = \overline{g}(t,r) + \int_0^1 H(t,s)e^{-s}\left(\overline{u}(s,r)\right)^2 ds, \tag{9.14}$$

where

$$g(t,r) = (\underline{g}(t,r), \overline{g}(t,r)) = (\underline{g_1}(t,r) + \underline{g_2}(t,r), \overline{g_1}(t,r) + \overline{g_2}(t,r)).$$

The exact solution of this problem is not known. This problem has been solved by the Bernstein polynomial collocation method and the Legendre wavelet method, and compared with the approximate-analytical method like the homotopy analysis method [178, 2]. Since the exact solution of this problem is unknown, the solution in HAM has been considered as the standard solution.

- ***Comparison with the HAM solution***

In the homotopy analysis method [178, 2], the mth order deformation equation approximating $\underline{u}(t,r)$ is given by

$$\begin{aligned} L\left[\underline{u}_m(t,r) - \chi_m \underline{u}_{m-1}(t,r)\right] &= \hbar \Re_m\left(\underline{u}_0, \underline{u}_1, ..., \underline{u}_{m-1}\right) \\ &= \hbar \frac{1}{(m-1)!} \frac{\partial^{m-1}}{\partial q^{m-1}} N(\varphi(t,r;q))|_{q=0} \end{aligned}$$

where

$$N(\varphi(t,r;q)) = \varphi(t,r;q) - \underline{g}(t,r) - \int_0^1 H(t,s)e^{-s}(\varphi(s,r;q))^2 ds$$

and $\chi_m = \begin{cases} 0, & m \leq 1, \\ 1, & m > 1. \end{cases}$

Here the auxiliary parameter [2] $\hbar = -1$ belongs to the convergence region of the HAM series solution.

Now using the mth order deformation equation of $\underline{u}(t,r)$, we recursively obtain

$$\begin{aligned} &\underline{u}_0(t,r) = 0, \\ &\underline{u}_1(t,r) = \frac{1}{10}(9 + r + 9t + rt - 47t^2 + 20et^2 - 3rt^2 + 28t^3 - 10et^3 + 2rt^3), \\ &\underline{u}_2(t,r) = \frac{1}{25}e^{-t}r^2t^6 - \frac{2}{5}e^{1-t}rt^6 + ..., \end{aligned}$$

and so on.

Thus the HAM solution

$$\underline{U}_m(t,r) = \sum_{i=0}^{m} \underline{u}_i(t,r).$$

Similarly, using the mth order deformation equation of $\overline{u}(t,r)$, we recursively obtain

$$\begin{aligned} &\overline{u}_0(t,r) = 0, \\ &\overline{u}_1(t,r) = \frac{1}{10}(11 - r + 11t - rt - 53t^2 + 20et^2 + 3rt^2 + 32t^3 - 10et^3 - 2rt^3), \\ &\overline{u}_2(t,r) = \frac{1}{100}e^{-1-t}\left(4er^2t^6 + 40e^2rt^6 - 128ert^6 + ...\right), \end{aligned}$$

and so on.

Thus the HAM solution

$$\overline{U}_m(t,r) = \sum_{i=0}^{m} \overline{u}_i(t,r).$$

TABLE 9.1: Numerical solutions for Example 9.3.1

x	t	$\underline{u}(t,r)$ HAM	$\underline{u}(t,r)$ BPCM	$\underline{u}(t,r)$ LWM	$\overline{u}(t,r)$ HAM	$\overline{u}(t,r)$ BPCM	$\overline{u}(t,r)$ LWM
0.25	0.2	1.13904	1.13903	1.13926	1.30376	1.30375	1.30403
	0.4	1.41286	1.41289	1.41283	1.57078	1.57081	1.57075
	0.6	1.75035	1.75035	1.75035	1.89387	1.89387	1.89386
	0.8	2.15758	2.15761	2.15785	2.29349	2.29352	2.29381
0.5	0.2	1.16649	1.16648	1.16672	1.27631	1.2763	1.27657
	0.4	1.43917	1.4392	1.43914	1.54445	1.54449	1.54443
	0.6	1.77427	1.77427	1.77426	1.86994	1.86994	1.86993
	0.8	2.18023	2.18026	2.18051	2.27084	2.27087	2.27115
0.75	0.2	1.19394	1.19393	1.19418	1.24885	1.24884	1.24910
	0.4	1.46549	1.46552	1.46546	1.51813	1.51816	1.51810
	0.6	1.79818	1.79818	1.79817	1.84602	1.84602	1.84601
	0.8	2.20288	2.20291	2.20317	2.24819	2.24821	2.24849

TABLE 9.2: Error analysis for Example 9.3.1 with regard to HAM

Error $0 \le t, r \le 1$	$\underline{u}(t,r)$ BPCM	$\underline{u}(t,r)$ LWM	$\overline{u}(t,r)$ BPCM	$\overline{u}(t,r)$ LWM
L_∞	0.3E-4	0.29E-3	0.4E-4	0.32E-3
L_2	0.754983E-4	0.600999E-3	0.768115E-4	0.690145E-3

The numerical results obtained by the Bernstein polynomial collocation method (BPCM) for $n_1 = n_2 = 4$ and the Legendre wavelet method (LWM) for $M_1 = M_2 = 4, k_1 = k_2 = 1$ have been compared with the results obtained by the 2nd order homotopy analysis method (HAM) of $u(t,r) = (\underline{u}(t,r), \overline{u}(t,r))$ for $r = 0.25,\ 0.5,\ 0.75$. These results have been shown in Tables 9.1-9.2. From Table 9.2, it may be easily observed that L_∞ and L_2 errors for the Bernstein polynomial collocation method are better than other method solutions.

Example 9.3.2. *Consider the following nonlinear Hammerstein fuzzy integral equation*

$$u(t) = g(t) + \int_0^1 \frac{ts}{3}(u(s))^2 ds, \quad t \in [0,1], \tag{9.15}$$

where

$$g(t,r) = \left[\frac{11}{12}t - \frac{1}{6}(1-r), \quad \frac{11}{12}t + \frac{1}{6}(1-r)\right], \quad t, r \in [0,1].$$

Eq. (9.15), *in the crisp sense, converted into a system as*

$$\underline{u}(t,r) = \frac{11}{12}t - \frac{1}{6}(1-r) + \int_0^1 \frac{ts}{3}\left(\underline{u}(s,r)\right)^2 ds \tag{9.16}$$

$$\overline{u}(t,r) = \frac{11}{12}t + \frac{1}{6}(1-r) + \int_0^1 \frac{ts}{3}\left(\overline{u}(s,r)\right)^2 ds \tag{9.17}$$

The exact solution of this problem is not known. This problem has been solved by the Bernstein polynomial collocation method and the Legendre wavelet method, and compared with the approximate-analytical method like the homotopy analysis method [178, 2]. Since the exact solution of this problem is unknown, the solution in HAM has been considered as the standard solution.

- ***Comparison with the HAM solution***

In the homotopy analysis method [178, 2], the mth order deformation equation approximating $\underline{u}(t,r)$ is given by

$$\begin{aligned} L\left[\underline{u}_m(t,r) - \chi_m \underline{u}_{m-1}(t,r)\right] &= \hbar \Re_m\left(\underline{u}_0, \underline{u}_1, \ldots, \underline{u}_{m-1}\right) \\ &= \hbar \frac{1}{(m-1)!}\frac{\partial^{m-1}}{\partial q^{m-1}} N(\varphi(t,r;q))|_{q=0}, \end{aligned}$$

where

$$N(\varphi(t,r;q)) = \varphi(t,r;q) - \frac{11}{12}t + \frac{1}{6}(1-r) - \int_0^1 \frac{ts}{3}(\varphi(s,r;q))^2 ds,$$

and $\chi_m = \begin{cases} 0, & m \leq 1, \\ 1, & m > 1. \end{cases}$

Here auxiliary parameter [2] $\hbar = -1$ belongs to the convergence region of the HAM series solution.

Now using the mth order deformation equation of $\underline{u}(t,r)$, we recursively obtain

$$\begin{aligned} \underline{u}_0(t,r) &= 0, \\ \underline{u}_1(t,r) &= -\frac{1}{6} + \frac{r}{6} + \frac{11t}{12}, \\ \underline{u}_2(t,r) &= \frac{211t}{5184} + \frac{2rt}{81} + \frac{r^2 t}{216}, \end{aligned}$$

and so on.

Thus the HAM solution

$$\underline{U}_m(t,r) = \sum_{i=0}^{m} \underline{u}_i(t,r)$$

Similarly, using the mth order deformation equation of $\overline{u}(t,r)$, we recursively

TABLE 9.3: Numerical solutions for Example 9.3.2

x	t	$\underline{u}(t,r)$ HAM	$\underline{u}(t,r)$ BPCM	$\underline{u}(t,r)$ LWM	$\overline{u}(t,r)$ HAM	$\overline{u}(t,r)$ BPCM	$\overline{u}(t,r)$ LWM
0.25	0.2	0.06780	0.06917	0.06917	0.32811	0.33258	0.33257
	0.4	0.26061	0.26334	0.26334	0.53122	0.54015	0.54015
	0.6	0.45341	0.45751	0.45751	0.73434	0.74772	0.74772
	0.8	0.64621	0.65168	0.65168	0.93745	0.95530	0.95529
0.5	0.2	0.11089	0.1126	0.11260	0.28443	0.28817	0.28818
	0.4	0.30511	0.30853	0.30853	0.48552	0.49301	0.49302
	0.6	0.49934	0.50446	0.50446	0.68662	0.69785	0.69786
	0.8	0.69356	0.70039	0.70040	0.88771	0.90269	0.90270
0.75	0.2	0.15410	0.15621	0.15621	0.24086	0.24399	0.24399
	0.4	0.34986	0.35408	0.35408	0.44006	0.44631	0.44631
	0.6	0.54562	0.55195	0.55195	0.63926	0.64864	0.64863
	0.8	0.74138	0.74982	0.74983	0.83845	0.85096	0.85095

TABLE 9.4: Error analysis for Example 9.3.2 with regard to HAM

Error $0 \leq t, r \leq 1$	$\underline{u}(t,r)$ BPCM	$\underline{u}(t,r)$ LWM	$\overline{u}(t,r)$ BPCM	$\overline{u}(t,r)$ LWM
L_∞	0.844E-2	0.845E-2	0.1785E-1	0.1784E-1
L_2	0.166465E-1	0.166393E-1	0.362121E-1	0.362082E-1

obtain

$$\begin{aligned}
\overline{u}_0(t,r) &= 0,\\
\overline{u}_1(t,r) &= \frac{1}{6} - \frac{r}{6} + \frac{11t}{12},\\
\overline{u}_2(t,r) &= \frac{563t}{5184} - \frac{7rt}{162} + \frac{r^2t}{216},\\
&\textit{and so on.}
\end{aligned}$$

Thus the HAM solution

$$\overline{U}_m(t,r) = \sum_{i=0}^{m} \overline{u}_i(t,r).$$

The numerical results obtained by the Bernstein polynomial collocation method (BPCM) for $n_1 = n_2 = 2$ *and the Legendre wavelet method (LWM) for* $M_1 = M_2 = 4, k_1 = k_2 = 1$ *have been compared with the results obtained by the* 3rd *order homotopy analysis method (HAM) of* $u(t,r) = (\underline{u}(t,r), \overline{u}(t,r))$ *for* $r = 0.25,\ 0.5,\ 0.75$. *These results have been shown in Tables 9.3-9.4. From Table 9.4, it may be easily observed that* L_∞ *and* L_2 *errors for the Bernstein polynomial collocation method are better than other method solutions.*

Example 9.3.3. *Consider the following nonlinear Hammerstein fuzzy integral equation*

$$u(t) = g(t) + \int_0^1 ts(u(s))^3 ds, \quad t \in [0,1], \tag{9.18}$$

where

$$g(t,r) = g_1(t,r) + g_2(t,r), \quad t,r \in [0,1],$$

where

$$g_1(t,r) = [2+r, 3-r],$$
$$g_2(t,r) = \left[-t\left(\frac{87}{10} + \frac{47}{4}r + 4r^2 + \frac{1}{2}r^3\right), -t\left(\frac{49}{10} - \frac{9}{2}r + \frac{5}{2}r^2 - \frac{1}{2}r^3\right)\right].$$

The exact solution of this problem is $u(t,r) = [2+t+r, \ 3-2t-r]$. *Eq.* (9.18), *in the crisp sense, converted into a system as*

$$\underline{u}(t,r) = \underline{g}(t,r) + \int_0^1 ts\left(\underline{u}(s,r)\right)^3 ds \tag{9.19}$$

$$\overline{u}(t,r) = \overline{g}(t,r) + \int_0^1 ts\left(\overline{u}(s,r)\right)^3 ds \tag{9.20}$$

where

$$g(t,r) = (\underline{g}(t,r), \overline{g}(t,r)) = (\underline{g_1}(t,r) + \underline{g_2}(t,r), \overline{g_1}(t,r) + \overline{g_2}(t,r))$$

Solving eq. (9.19) *by the Bernstein polynomial collocation method for* $n_1 = n_2 = 1$, *we obtain the unknowns as*

$$c_{0,0} = 2, \quad c_{0,1} = 3, \quad c_{1,0} = 3, \quad c_{1,1} = 4$$

and the solution can be obtained as $\underline{u}(t,r) = \sum_{i=0}^{1}\sum_{j=0}^{1} c_{i,j} B_{i,1}(t) B_{j,1}(r) = 2+t+r$.
Similarly, solving eq. (9.20) *by the* 1st *order Bernstein polynomial collocation method, we obtain the unknowns as*

$$c_{0,0} = 3, \quad c_{0,1} = 2, \quad c_{1,0} = 1, \quad c_{1,1} = 0$$

and the solution can be obtained as $\overline{u}(t,r) = \sum_{i=0}^{1}\sum_{j=0}^{1} c_{i,j} B_{i,1}(t) B_{j,1}(r) = 3-2t-r$.

Solving eq. (9.19) *by the Legendre wavelet method for* $M_1 = M_2 = 2, k_1 = k_2 = 1$, *we obtain the unknowns as*

$$C^T = [3.0003, 0.28862, 0.28885, -0.0000332746],$$

and the solution can be obtained as $\underline{u}(t,r) = C^T \Psi(t,r) = 2 + r + t - 0.000399295rt$.

Similarly, solving eq. (9.20) *by the Legendre wavelet method for* $M_1 = M_2 = 2, k_1 = k_2 = 1$, *we obtain the unknowns as*

$$C^T = [1.52366, -0.216966, -0.563689, 0.0414012],$$

and the solution can be obtained as $\overline{u}(t,r) = C^T\Psi(t,r) = 3 - r - 2.20108t + 0.496815rt$.

9.4 Numerical solution of Hammerstein-Volterra fuzzy delay integral equation

In this section, we consider the nonlinear Hammerstein-Volterra fuzzy delay integral equation of the form

$$x(t) = \begin{cases} g(t) + \int_{t-\tau}^{t} H(t,s)F(s,x(s))ds, & t \in [0,T], T > 0, \\ \Phi(t), & t \in [-\tau, 0], \tau > 0, \end{cases} \tag{9.21}$$

where $x(t)$, $g(t)$ and $\Phi(t)$ are fuzzy valued functions, and the sign of $H(t,s)$ does not change in $[0,1]$. Let

$$\begin{aligned} x(t,r) &= (\underline{x}(t,r), \overline{x}(t,r)), \\ g(t,r) &= (\underline{g}(t,r), \overline{g}(t,r)), \\ \Phi(t,r) &= (\underline{\Phi}(t,r), \overline{\Phi}(t,r)). \end{aligned}$$

Eq. (9.21), in the crisp sense, converted into a system as

$$\underline{x}(t,r) = \begin{cases} \underline{g}(t,r) + \int_{t-\tau}^{t} H(t,s)F(s,\underline{x}(s,r))ds, & r \in [0,1], \quad t \in [0,T] \\ \underline{\Phi}(t,r), & r \in [0,1], \quad t \in [-\tau, 0], \end{cases} \tag{9.22}$$

$$\overline{x}(t,r) = \begin{cases} \overline{g}(t,r) + \int_{t-\tau}^{t} H(t,s)F(s,\overline{x}(s,r))ds, & r \in [0,1], \quad t \in [0,T] \\ \overline{\Phi}(t,r), & r \in [0,1], \quad t \in [-\tau, 0]. \end{cases} \tag{9.23}$$

9.4.1 Numerical scheme for Hammerstein-Volterra fuzzy delay integral equation

Let us consider the nonlinear fuzzy Hammerstein-Volterra delay integral equation given in equation (9.21) and we approximate the unknown function

$x(t,r) \in [-\tau, T] \times [0,1]$ by the two dimensional Bernoulli wavelet method (See Section 2.6 of Chapter 2) after using transformation $u = \frac{t+\tau}{T+\tau}$. Now the eq. (9.21) can be reduced as

$$C^T\Psi(t,r) = \begin{cases} g(t,r) + \int_{t-\tau}^{t} H(t,s)F[C^T\Psi(s,r)]ds, & t \in [0,T], \\ \Phi(t,r), & t \in [-\tau, 0]. \end{cases} \tag{9.24}$$

Putting the collocation points

$$t_i = \frac{(2i-1)(T+\tau)}{M_1 2^{k_1}} - \tau, \qquad i = 1,2,...,M_1 2^{k_1-1}$$

and

$$r_j = \frac{2j-1}{M_2 2^{k_2}}, \qquad j = 1,2,...,M_2 2^{k_2-1}$$

in eq. (9.24), we have

$$C^T\Psi(t_i,r_j) = \begin{cases} g(t_i,r_j) + \int_{t_i-\tau}^{t_i} H(t_i,s)F[C^T\Psi(s,r_j)]ds, & 0 \le t_i < T, \\ \Phi(t_i,r_j), & \tau < t_i < 0. \end{cases} \tag{9.25}$$

For more simplification, the Guass-Legendre quadrature rule has been applied for approximating the integration in eq. (9.25). In order to apply the Guass-Legendre quadrature rule, we convert the limit of integration $[t_i - \tau, t_i]$ to $[-1,1]$ by transformation $\varphi = \frac{2s-2t_i+\tau}{\tau}$. Thus we obtain

$$C^T\Psi(t_i,r_j) = \begin{cases} g(t_i,r_j) + \frac{\tau}{2}\int_{-1}^{1} H\left(t_i, \frac{\tau(\varphi-1)+2t_i}{2}\right) F\left[C^T\Psi\left(\frac{\tau(\varphi-1)+2t_i}{2}, r_j\right)\right] d\varphi, & 0 \le t_i < T, \\ \Phi(t_i,r_j), & -\tau < t_i < 0. \end{cases} \tag{9.26}$$

Applying the Gauss-Legendre quadrature rule to the eq. (9.26), we have

$$C^T\Psi(t_i,r_j) = \begin{cases} g(t_i,r_j) + \frac{\tau}{2}\sum_{l=1}^{p} w_l H\left(t_i, \frac{\tau(\varphi_l-1)+2t_i}{2}\right) F\left[C^T\Psi\left(\frac{\tau(\varphi_l-1)+2t_i}{2}, r_j\right)\right], & 0 \le t_i < T, \\ \Phi(t_i,r_j), & -\tau < t_i < 0, \end{cases} \tag{9.27}$$

where φ_l, $l = 1,...,p$ are Legendre-Gauss points, i.e., the roots of pth order Legendre polynomials and w_l are the corresponding weights.

Eq. (9.27) gives a system of $(2^{k-1}M)^2$ number of algebraic equations with the same number of unknowns for C. After solving this system numerically, we can obtain the value of c_{n_1,m_1,n_2,m_2} for $n_1 = 1,2,...,2^{k_1-1}$, $m_1 = 0,1,...,M_1$, $n_2 = 1,2,...,2^{k_2-1}$, $m_2 = 0,1,...,M_2$ and hence obtain the solution for $x(t,r)$.

9.4.2 Convergence analysis and error estimate

Theorem 9.4.1. *If $f(u,v) \in L^2(R \times R)$ is a continuous function defined on $[0,1] \times [0,1]$ and $|f(u,v)| \le K$, then the Bernoulli wavelets expansion of $f(u,v)$ defined in eq. (2.65) of Chapter 2 converges uniformly and also*

$$|c_{n_1,m_1,n_2,m_2}| < K \frac{A_1 A_2}{2^{\frac{k_1-1}{2}} 2^{\frac{k_2-1}{2}}} \frac{16 m_1!}{(2\pi)^{m_1+1}} \frac{16 m_2!}{(2\pi)^{m_2+1}},$$

where $A_1 = \dfrac{1}{\sqrt{\frac{(-1)^{m_1-1}(m_1!)^2}{(2m_1)!}\alpha_{2m_1}}}$ *and* $A_2 = \dfrac{1}{\sqrt{\frac{(-1)^{m_2-1}(m_2!)^2}{(2m_2)!}\alpha_{2m_2}}}$.

Proof. Any function $f(u,v) \in L^2([0,1] \times [0,1])$ can be expressed as the two dimensional Bernoulli wavelets as

$$\begin{aligned} f(u,v) &= \sum_{n_1=1}^{2^{k_1-1}} \sum_{m_1=0}^{M_1-1} \sum_{n_2=1}^{2^{k_2-1}} \sum_{m_2=0}^{M_2-1} c_{n_1,m_1,n_2,m_2} \psi_{n_1,m_1,n_2,m_2}(u,v) \\ &= \sum_{n_1=1}^{2^{k_1-1}} \sum_{m_1=0}^{M_1-1} \sum_{n_2=1}^{2^{k_2-1}} \sum_{m_2=0}^{M_2-1} c_{n_1,m_1,n_2,m_2} \psi_{n_1,m_1}(u) \psi_{n_2,m_2}(v), \end{aligned}$$

where the coefficients c_{n_1,m_1,n_2,m_2} can be determined as

$$c_{n_1,m_1,n_2,m_2} = \langle \langle f(u,v), \psi_{n_1,m_1}(u) \rangle, \psi_{n_2,m_2}(v) \rangle.$$

Now for $m_1, \ m_2 > 0$,

$$\begin{aligned} c_{n_1,m_1,n_2,m_2} &= \langle \langle f(u,v), \psi_{n_1,m_1}(u) \rangle, \psi_{n_2,m_2}(v) \rangle \\ &= \int_0^1 \left[\int_0^1 f(u,v) \psi_{n_1,m_1}(u) du \right] \psi_{n_2,m_2}(v) dv \\ &= \int_{I_{n_2 k_2}} \left[\int_{I_{n_1 k_1}} f(u,v) \psi_{n_1,m_1}(u) du \right] \psi_{n_2,m_2}(v) dv \\ &= 2^{\frac{k_1-1}{2}} 2^{\frac{k_2-1}{2}} A_1 A_2 \int_{I_{n_2 k_2}} \left[\int_{I_{n_1 k_1}} f(u,v) \beta_{m_1}(2^{k_1-1}u - n_1 + 1) du \right] \\ &\quad \times \beta_{m_2}(2^{k_2-1}v - n_2 + 1) dv, \end{aligned}$$

where $I_{n_1 k_1} = [\frac{n_1-1}{2^{k_1-1}}, \frac{n_1}{2^{k_1-1}}]$, $I_{n_2 k_2} = [\frac{n_2-1}{2^{k_2-1}}, \frac{n_2}{2^{k_2-1}}]$.
Now, changing the variable $2^{k_1-1}u - n_1 + 1 = t$, we have

$$c_{n_1,m_1,n_2,m_2} = \frac{1}{2^{\frac{k_1-1}{2}}} 2^{\frac{k_2-1}{2}} A_1 A_2 \int_{I_{n_2 k_2}} \left[\int_0^1 f\left(\frac{t + n_1 - 1}{2^{k_1-1}}, v \right) \beta_{m_1}(t) dt \right] \times \beta_{m_2}(2^{k_2-1}v - n_2 + 1) dv.$$

Similarly, changing the variable for v as $2^{k_2-1}v - n_2 + 1 = \tilde{t}$, we have

$$c_{n_1,m_1,n_2,m_2} = \frac{A_1}{2^{\frac{k_1-1}{2}}} \frac{A_2}{2^{\frac{k_2-1}{2}}} \int_0^1 \left[\int_0^1 f\left(\frac{t + n_1 - 1}{2^{k_1-1}}, \frac{\tilde{t} + n_2 - 1}{2^{k_2-1}} \right) \beta_{m_1}(t) dt \right] \beta_{m_2}(\tilde{t}) d\tilde{t}.$$

Now

$$
\begin{aligned}
|c_{n_1,m_1,n_2,m_2}| &\leq \frac{A_1}{2^{\frac{k_1-1}{2}}}\frac{A_2}{2^{\frac{k_2-1}{2}}}\int_0^1 \\
&\times \left[\int_0^1 \left|f\left(\frac{t+n_1-1}{2^{k_1-1}}, \frac{\tilde{t}+n_2-1}{2^{k_2-1}}\right)\right| |\beta_{m_1}(t)|dt\right] |\beta_{m_2}(\tilde{t})|d\tilde{t} \\
&\leq \frac{A_1}{2^{\frac{k_1-1}{2}}}\frac{A_2}{2^{\frac{k_2-1}{2}}}K\left(\int_0^1 |\beta_{m_1}(t)|dt\right)\left(\int_0^1 |\beta_{m_2}(\tilde{t})|d\tilde{t}\right) \\
&\leq \frac{A_1}{2^{\frac{k_1-1}{2}}}\frac{A_2}{2^{\frac{k_2-1}{2}}}K\frac{16m_1!}{(2\pi)^{m_1+1}}\frac{16m_2!}{(2\pi)^{m_2+1}},
\end{aligned}
$$

using the property of Bernoulli polynomials.

This means that the series $\sum_{n_1=1}^{2^{k_1-1}}\sum_{m_1=0}^{M_1-1}\sum_{n_2=1}^{2^{k_2-1}}\sum_{m_2=0}^{M_2-1} c_{n_1,m_1,n_2,m_2}$ is absolutely convergent and hence the series

$$
\sum_{n_1=1}^{2^{k_1-1}}\sum_{m_1=0}^{M_1-1}\sum_{n_2=1}^{2^{k_2-1}}\sum_{m_2=0}^{M_2-1} c_{n_1,m_1,n_2,m_2}\psi_{n_1,m_1,n_2,m_2}(u,v)
$$

is uniformly convergent [115]. □

Theorem 9.4.2. *Let* $f^*(t,r) = \sum_{n_1=1}^{2^{k_1-1}}\sum_{m_1=0}^{M_1-1}\sum_{n_2=1}^{2^{k_2-1}}\sum_{m_2=0}^{M_2-1} c_{n_1,m_1,n_2,m_2}\psi_{n_1,m_1,n_2,m_2}(t,r)$ *be the truncated series; then the truncated error* $E_{n_1,m_1,n_2,m_2}(t,r)$ *can be defined as*

$$
\begin{aligned}
\|E_{n_1,m_1,n_2,m_2}(t,r)\|_2^2 \leq & \sum_{n_1=2^{k_1-1}+1}^{\infty}\sum_{m_1=M_1}^{\infty}\sum_{n_2=2^{k_2-1}+1}^{\infty}\sum_{m_2=M_2}^{\infty} \\
& \times \left(\frac{A_1}{2^{\frac{k_1-1}{2}}}\frac{A_2}{2^{\frac{k_2-1}{2}}}\frac{16m_1!}{(2\pi)^{m_1+1}}\frac{16m_2!}{(2\pi)^{m_2+1}}K\right)^2
\end{aligned}
$$

by using Theorem 9.4.1.

Proof. Any function $f(t,r) \in L^2([0,1]\times[0,1])$ can be expressed by the Bernoulli wavelets as

$$
f(t,r) = \sum_{n_1=1}^{\infty}\sum_{m_1=0}^{\infty}\sum_{n_2=1}^{\infty}\sum_{m_2=0}^{\infty} c_{n_1,m_1,n_2,m_2}\psi_{n_1,m_1,n_2,m_2}(t,r).
$$

If $f^*(t,r)$ is the truncated series, then the truncated error term can be calculated as

$$
\begin{aligned}
E_{n_1,m_1,n_2,m_2}(t,r) &= f(t,r) - f^*(t,r) \\
&= \sum_{n_1=2^{k_1-1}+1}^{\infty}\sum_{m_1=M_1}^{\infty}\sum_{n_2=2^{k_2-1}+1}^{\infty}\sum_{m_2=M_2}^{\infty} c_{n_1,m_1,n_2,m_2}\psi_{n_1,m_1,n_2,m_2}(t,r).
\end{aligned}
$$

Now,

$$\|E_{n_1,m_1,n_2,m_2}(t,r)\|_2^2$$

$$= \| \sum_{n_1=2^{k_1-1}+1}^{\infty} \sum_{m_1=M_1}^{\infty} \sum_{n_2=2^{k_2-1}+1}^{\infty} \sum_{m_2=M_2}^{\infty} c_{n_1,m_1,n_2,m_2}\psi_{n_1,m_1,n_2,m_2}(t,r)\|_2^2$$

$$= \int_0^1 \int_0^1 | \sum_{n_1=2^{k_1-1}+1}^{\infty} \sum_{m_1=M_1}^{\infty} \sum_{n_2=2^{k_2-1}+1}^{\infty} \sum_{m_2=M_2}^{\infty} c_{n_1,m_1,n_2,m_2}\psi_{n_1,m_1,n_2,m_2}(t,r)|^2 dtdr$$

$$\leq \sum_{n_1=2^{k_1-1}+1}^{\infty} \sum_{m_1=M_1}^{\infty} \sum_{n_2=2^{k_2-1}+1}^{\infty} \sum_{m_2=M_2}^{\infty} |c_{n_1,m_1,n_2,m_2}|^2 \int_0^1 \int_0^1 |\psi_{n_1,m_1,n_2,m_2}(t,r)|^2 dtdr$$

$$\leq \sum_{n_1=2^{k_1-1}+1}^{\infty} \sum_{m_1=M_1}^{\infty} \sum_{n_2=2^{k_2-1}+1}^{\infty} \sum_{m_2=M_2}^{\infty} \left(\frac{A_1}{2^{\frac{k_1-1}{2}}} \frac{A_2}{2^{\frac{k_2-1}{2}}} \frac{16m_1!}{(2\pi)^{m_1+1}} \frac{16m_2!}{(2\pi)^{m_2+1}} K\right)^2.$$

□

Theorem 9.4.3. *Let $x^*(t,r)$ be the approximate solution of the equation*

$$x(t,r) = \begin{cases} g(t,r) + \int_{t-\tau}^{t} H(t,s)F(s,x(s,r))ds, & t \in [0,T], r \in [0,1], \\ \Phi(t,r), & t \in [-\tau,0], r \in [0,1], \end{cases} \tag{9.28}$$

where $\|H(t,s)\|_\infty \leq M_H$ and F is a lipschitz continuous function; then the error term $R(t,r)$ can be estimated as

$$\|R(t,r)\|_\infty \leq \frac{\tau}{2} M_H L \sum_{j=1}^{n} w_j \left\| E_{n_1,m_1,n_2,m_2}\left(\frac{\tau}{2}(z_j-1)+t,r\right)\right\|_\infty$$
$$+ \|{}_1E_{GL}(t,\xi,r) -{}_2 E_{GL}(t,\xi,r)\|_\infty + \|E_{n_1,m_1,n_2,m_2}(t,r)\|_\infty.$$

Proof. Let $x^*(t,r)$ be the truncated series which approximates the unknown function of equation (9.28); then

$$x^*(t,r) = \begin{cases} g(t,r) + \int_{t-\tau}^{t} H(t,s)F(s,x^*(s,r))ds, & t \in [0,T], r \in [0,1] \\ \Phi(t,r), & t \in [-\tau,0], r \in [0,1], \end{cases} + R(t,r), \tag{9.29}$$

where $R(t,r)$ is the residual. In order to apply Gauss-quadrature, transform the limit of integration from $[t-\tau,t]$ to $[-1,1]$ as $s = \frac{\tau}{2}(z-1)+t$.

$$x^*(t,r) = R(t,r)$$
$$+ \begin{cases} g(t,r) + \frac{\tau}{2} \int_{-1}^{1} H\left(t, \frac{\tau}{2}(z-1)+t\right) \times \\ \quad F\left(\frac{\tau}{2}(z-1)+t, x^*\left(\frac{\tau}{2}(z-1)+t,r\right)\right) dz, & t \in [0,T], r \in [0,1], \\ \Phi(t,r), & t \in [-\tau,0], r \in [0,1]. \end{cases} \tag{9.30}$$

Put $U(t,z,r) = H\left(t, \frac{\tau}{2}(z-1)+t\right) F\left(\frac{\tau}{2}(z-1)+t, x^*\left(\frac{\tau}{2}(z-1)+t, r\right)\right)$ and applying the Gauss-quadrature rule in eq. (9.30), we have

$$x^*(t,r) = R(t,r) + \begin{cases} g(t,r) + \frac{\tau}{2}\sum_{j=1}^{n} U(t,z_j,r) +_2 E_{GL}(t,\xi,r), & t \in [0,T], r \in [0,1], \\ \Phi(t,r), & t \in [-\tau,0], r \in [0,1], \end{cases} \tag{9.31}$$

where ${}_2E_{GL}(t,\xi,r)$ is the error term in the Gauss-Legendre quadrature rule and obtained as

$${}_2E_{GL}(t,\xi,r) = \frac{2^{2n+1}(n!)^4}{(2n+1)[(2n)!]^3} U^{(2n)}(t,\xi,r). \tag{9.32}$$

Similarly, we can write

$$x(t,r) = \begin{cases} g(t,r) + \frac{\tau}{2}\sum_{j=1}^{n} V(t,z_j,r) +_1 E_{GL}(t,\xi,r), & t \in [0,T], r \in [0,1], \\ \Phi(t,r), & t \in [-\tau,0], r \in [0,1], \end{cases} \tag{9.33}$$

where

$${}_1E_{GL}(t,\xi,r) = \frac{2^{2n+1}(n!)^4}{(2n+1)[(2n)!]^3} V^{(2n)}(t,\xi,r). \tag{9.34}$$

Now, from eq. (9.31) and eq. (9.33), we get

$$\begin{aligned} x(t,r) - x^*(t,r) = \frac{\tau}{2}\sum_{j=1}^{n} w_j(V(t,z_j,r) - U(t,z_j,r)) \\ + ({}_1E_{GL}(t,\xi,r) -_2 E_{GL}(t,\xi,r)) - R(t,r) \end{aligned} \tag{9.35}$$

$$\begin{aligned} R(t,r) = \frac{\tau}{2}\sum_{j=1}^{n} w_j(V(t,z_j,r) - U(t,z_j,r)) \\ + ({}_1E_{GL}(t,\xi,r) -_2 E_{GL}(t,\xi,r)) - E_{n_1,m_1,n_2,m_2}(t,r) \end{aligned}$$

Therefore,

$$\begin{aligned} \|R(t,r)\|_\infty \le \frac{\tau}{2}\sum_{j=1}^{n} w_j \|V(t,z_j,r) - U(t,z_j,r)\|_\infty \\ + \|{}_1E_{GL}(t,\xi,r) -_2 E_{GL}(t,\xi,r)\|_\infty \\ + \|E_{n_1,m_1,n_2,m_2}(t,r)\|_\infty. \end{aligned} \tag{9.36}$$

Assuming that $\|H(t,s)\|_\infty \le M_H$ and the nonlinear term $F(.,.)$ satisfies the Lipschitz condition, we have

$$\begin{aligned}
&\|V(t,z_j,r)-U(t,z_j,r)\|_\infty \\
&\leq \left\|H\left(t,\frac{\tau}{2}(z_j-1)+t\right)\right\|_\infty \|F\left(\frac{\tau}{2}(z_j-1)+t, x\left(\frac{\tau}{2}(z_j-1)+t,r\right)\right) \\
&\quad - F\left(\frac{\tau}{2}(z_j-1)+t, x^*\left(\frac{\tau}{2}(z_j-1)+t,r\right)\right)\|_\infty \\
&\leq M_H L\left\|x\left(\frac{\tau}{2}(z_j-1)+t,r\right)-x^*\left(\frac{\tau}{2}(z_j-1)+t,r\right)\right\|_\infty \\
&= M_H L\left\|E_{n_1,m_1,n_2,m_2}\left(\frac{\tau}{2}(z_j-1)+t,r\right)\right\|_\infty,
\end{aligned} \tag{9.37}$$

where L is Lipschitz constant. Hence from eq. (9.36), we have

$$\begin{aligned}
\|R(t,r)\|_\infty \leq \frac{\tau}{2} M_H L \sum_{j=1}^{n} w_j \left\|E_{n_1,m_1,n_2,m_2}\left(\frac{\tau}{2}(z_j-1)+t,r\right)\right\|_\infty \\
+ \|{}_1E_{GL}(t,\xi,r) - {}_2E_{GL}(t,\xi,r)\|_\infty + \|E_{n_1,m_1,n_2,m_2}(t,r)\|_\infty.
\end{aligned} \tag{9.38}$$

□

9.4.3 Illustrative examples

In order to show the accuracy of the present method three problems have been solved by both the present method and the B-spline wavelet method. The same procedure has been implemented for B-spline wavelet method to solve these problems.

Example 9.4.1. *Let us consider the linear Volterra fuzzy delay integral equation [42]*

$$x(t) = \begin{cases} e^{t-\tau} + \int_{t-\tau}^{t} x(s)ds, & t \in [0, 0.5], \\ \Phi(t), & t \in [-\tau, 0], \end{cases}$$

with $\tau = 0.5$, $T = 0.5$ *and* $\Phi : [-0.5, 0] \to E$ *defined as*

$$\Phi(t,r) = [\underline{\Phi}(t,r), \overline{\Phi}(t,r)] = [e^t, e^t + 0.3(1-r)], \quad r \in [0,1], \quad t \in [-0.5, 0].$$

Similarly, $g : [0, 0.5] \to E$ *is given by*

$$g(t,r) = [\underline{g}(t,r), \overline{g}(t,r)] = [e^{t-\tau}, e^{t-\tau} + 0.3\tau(1-r)], \quad r \in [0,1], \quad t \in [0, 0.5].$$

The exact solution $x^* : [-0.5, 0.5] \to E$ *is given by*

$$x^*(t,r) = [\underline{x^*}(t,r), \overline{x^*}(t,r)] = [e^t, e^t + 0.3(1-r)], \quad r \in [0,1], \quad t \in [-0.5, 0.5].$$

TABLE 9.5: Comparison of numerical solutions for $\underline{x}(t, r)$ in Example 9.4.1

t	Bernoulli wavelet method			B-spline wavelet method		
	$\lvert e_{0.25}\rvert$	$\lvert e_{0.5}\rvert$	$\lvert e_{0.75}\rvert$	$\lvert e_{0.25}\rvert$	$\lvert e_{0.5}\rvert$	$\lvert e_{0.75}\rvert$
0.1	1.05E-5	1.05E-5	1.05E-5	5.39E-5	5.39E-5	5.39E-5
0.2	4.15E-6	4.15E-6	4.15E-6	6.74E-5	6.74E-5	6.74E-5
0.3	5.17E-6	5.17E-6	5.17E-6	4.03E-5	4.03E-5	4.03E-5
0.4	9.76E-6	9.76E-6	9.76E-6	4.88E-5	4.88E-5	4.88E-5
0.5	9.22E-5	9.22E-5	9.22E-5	2.07E-4	2.07E-4	2.07E-4

TABLE 9.6: Comparison of numerical solutions for $\overline{x}(t, r)$ in Example 9.4.1

t	Bernoulli wavelet method			B-spline wavelet method		
	$\lvert e_{0.25}\rvert$	$\lvert e_{0.5}\rvert$	$\lvert e_{0.75}\rvert$	$\lvert e_{0.25}\rvert$	$\lvert e_{0.5}\rvert$	$\lvert e_{0.75}\rvert$
0.1	1.16E-5	1.16E-5	1.16E-5	5.27E-5	5.27E-5	5.27E-5
0.2	3.05E-6	3.05E-6	3.05E-6	6.20E-5	6.20E-5	6.20E-5
0.3	5.18E-6	5.18E-6	5.18E-6	4.13E-5	4.13E-5	4.13E-5
0.4	9.68E-6	9.68E-6	9.68E-6	4.08E-5	4.08E-5	4.08E-5
0.5	9.32E-5	9.32E-5	9.32E-5	2.17E-4	2.17E-4	2.17E-4

The above problem has been solved by the Bernoulli wavelet method for $M_1 = M_2 = 4$, $k_1 = k_2 = 2$ *and then the results have been compared with that obtained by B-spline wavelet method for* $m = 4$. *Here we take* $r = 0.25, \ 0.5, \ 0.75$ *and calculate the absolute errors as* $|e_r| = |x(t, r) - x^*(t, r)|$. *This comparison has been presented in Table 9.5 and Table 9.6. The overall computational times for* $\underline{x}(t, r)$ *and* $\overline{x}(t, r)$ *are 1.327 seconds and 1.342 seconds respectively by the Bernoulli wavelet method. On the other hand, the overall computational times for* $\underline{x}(t, r)$ *and* $\overline{x}(t, r)$ *are 475.71 seconds and 461.42 seconds respectively by the B-spline wavelet method. Figure 9.1 shows the approximate solutions of* $\overline{x}(t, r)$ *for* $r = 0, 0.25, 0.5, 0.75, 1$ *and the approximate solution of* $\underline{x}(t, r)$ *independent of* r.

Example 9.4.2. *Let us consider the nonlinear Hammerstein-Volterra fuzzy delay integral equation [42]*

$$x(t) = \begin{cases} g(t) + \int_{t-\tau}^{t} \ln x(s) ds, & t \in [0, 1], \\ \Phi(t), & t \in [-\tau, 0], \end{cases}$$

with $\tau = 0.5$, $T = 1$ *and* $\Phi : [-0.5, 0] \to E$ *defined as*

$$\Phi(t, r) = [\underline{\Phi}(t, r), \overline{\Phi}(t, r)] = [e^{t+1}, e^{tr}], \quad r \in [0, 1], \quad t \in [-0.5, 0].$$

Similarly, $g : [0, 1] \to E$ *is given by*

$$g(t, r) = [\underline{g}(t, r), \overline{g}(t, r)] = [e^{t+1} - \tau - t\tau + \frac{\tau^2}{2}, e^{tr} + r\tau(0.25 - t)],$$
$$r \in [0, 1], \quad t \in [0, 1].$$

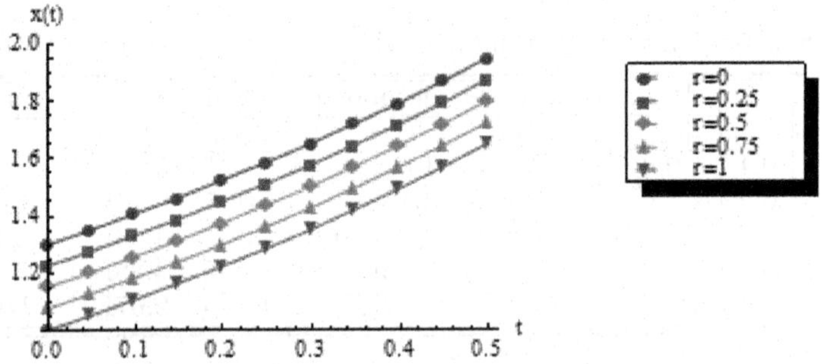

FIGURE 9.1: Approximate solution of $\overline{x}(t,r)$ for $r = 0, 0.25, 0.5, 0.75, 1$ of Example 9.4.1

TABLE 9.7: Comparison of numerical solutions for $\underline{x}(t, r)$ in Example 9.4.2

t	Bernoulli wavelet method			B-spline wavelet method		
	$\|e_{0.25}\|$	$\|e_{0.5}\|$	$\|e_{0.75}\|$	$\|e_{0.25}\|$	$\|e_{0.5}\|$	$\|e_{0.75}\|$
0.1	1.28E-4	1.28E-4	1.28E-4	2.07E-4	2.07E-4	2.07E-4
0.2	2.65E-4	2.65E-4	2.65E-4	1.01E-4	1.01E-4	1.01E-4
0.3	5.53E-4	5.53E-4	5.53E-4	1.43E-4	1.43E-4	1.43E-4
0.4	2.26E-4	2.26E-4	2.26E-4	3.60E-4	3.60E-4	3.60E-4
0.5	7.87E-5	7.87E-5	7.87E-5	4.68E-4	4.68E-4	4.68E-4
0.6	1.53E-4	1.53E-4	1.53E-4	4.30E-4	4.30E-4	4.30E-4
0.7	7.54E-5	7.54E-5	7.54E-5	2.05E-4	2.05E-4	2.05E-4
0.8	2.10E-4	2.10E-4	2.10E-4	2.44E-4	2.44E-4	2.44E-4
0.9	5.44E-5	5.44E-5	5.44E-5	9.73E-4	9.73E-4	9.73E-4
1.0	1.79E-3	1.79E-3	1.79E-3	2.04E-3	2.04E-3	2.04E-3

The exact solution $x^* : [-0.5, 1] \to E$ is given by

$$x^*(t,r) = [\underline{x}^*(t,r), \overline{x^*}(t,r)] = [e^{t+1}, e^{tr}], \quad r \in [0,1], \quad t \in [-0.5, 1].$$

The above problem has been solved by the Bernoulli wavelet method for $M_1 = M_2 = 4$, $k_1 = k_2 = 2$ and then the results have been compared with that obtained by B-spline wavelet method for $m = 4$. Here we take $r = 0.25$, 0.5, 0.75 and calculate the absolute errors as $|e_r| = |x(t,r) - x^*(t,r)|$. This comparison has been presented in Table 9.7 and Table 9.8. The overall computational times for $\underline{x}(t,r)$ and $\overline{x}(t,r)$ are 7.425 seconds and 7.55 seconds respectively by the Bernoulli wavelet method. On the other hand, the overall computational times for $\underline{x}(t,r)$ and $\overline{x}(t,r)$ are 6125.44 seconds and 6210.45 seconds respectively by the B-spline wavelet method. Figure 9.2 shows the approximate solutions of $\overline{x}(t,r)$ for $r = 0, 0.25, 0.5, 0.75, 1$ and the approximate solution of $\underline{x}(t,r)$ independent of r.

TABLE 9.8: Comparison of numerical solutions for $\overline{x}(t,r)$ in Example 9.4.2

t	Bernoulli wavelet method			B-spline wavelet method		
	$\lvert e_{0.25}\rvert$	$\lvert e_{0.5}\rvert$	$\lvert e_{0.75}\rvert$	$\lvert e_{0.25}\rvert$	$\lvert e_{0.5}\rvert$	$\lvert e_{0.75}\rvert$
0.1	1.97E-7	3.14E-6	1.59E-5	2.90E-6	1.57E-5	3.49E-5
0.2	3.68E-7	5.85E-6	2.92E-5	2.01E-6	7.70E-6	1.38E-5
0.3	6.35E-7	1.10E-5	6.52E-5	5.70E-6	1.36E-5	2.22E-5
0.4	1.78E-8	4.52E-7	1.87E-5	8.56E-6	3.32E-5	6.10E-5
0.5	3.92E-7	5.48E-6	4.18E-6	1.64E-5	4.18E-5	7.00E-5
0.6	1.06E-6	1.66E-5	2.14E-5	2.98E-5	3.78E-5	4.29E-5
0.7	1.79E-6	2.87E-5	1.68E-5	4.91E-5	1.95E-5	2.72E-5
0.8	2.77E-6	4.61E-5	9.89E-6	7.41E-5	1.20E-5	1.42E-4
0.9	4.68E-6	8.18E-5	3.07E-6	1.04E-4	5.69E-5	3.06E-4
1.0	8.66E-6	1.58E-4	1.75E-4	1.40E-4	1.16E-4	5.27E-4

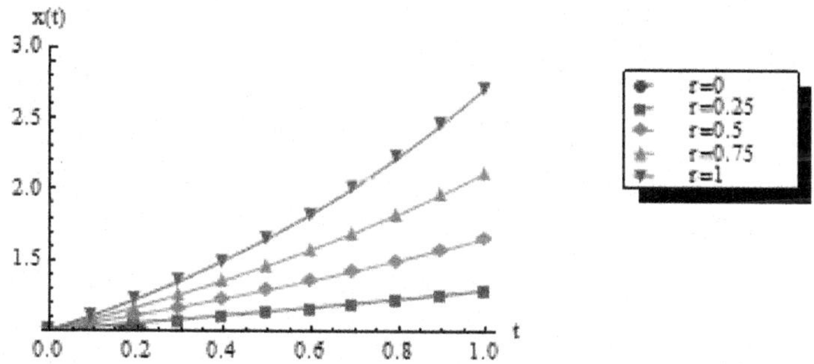

FIGURE 9.2: Approximate solution of $\overline{x}(t,r)$ for $r = 0, 0.25, 0.5, 0.75, 1$ of Example 9.4.2

Example 9.4.3. *According to the epidemic model presented in [42, 55], we consider $x(s)$ to be the proportion of infectious individuals at the moment s; $F(x(s))$ be the proportion of new infected cases on unit time, and $g(s)$ to be the proportion of immigrants that still have the disease at the moment s. Considering $P(s)$ as the probability of having the infection for a time at least s after infection, the spread of infection is governed by the integral equation*

$$x(t) = \begin{cases} g(t) + \int_{t-\tau}^{t} P(t-s)F(x(s))ds, & t \in [0,T], \\ \Phi(t), & t \in [-\tau, 0]. \end{cases}$$

We suppose that the proportion of new infected cases on unit time, $F(x(s))$, is proportional with $x(s)$ and P is a crisp positive decreasing function with $P(0) = 1$. Let $\tau = T = 0.5$, and $\Phi : [-0.5, 0] \to E$, $g : [0, 0.5] \to E$. Since it is natural to suppose that the proportion of immigrants that still have the disease is decreasing in time according to the decisions of the authorities, we have the following model

$$x(t) = \begin{cases} g(t) + \int_{t-\tau}^{t} e^{-(t-s)} x(s) ds, & t \in [0, 0.5], \\ \Phi(t), & t \in [-\tau, 0], \end{cases}$$

with

$$\Phi(t,r) = [\underline{\Phi}(t,r), \overline{\Phi}(t,r)] = [e^{-t} - 0.2(1-r), e^{-t} + 0.2(1-r)],$$
$$r \in [0,1], \quad t \in [-0.5, 0].$$

and

$$g(t,r) = [\underline{g}(t,r), \overline{g}(t,r)] = [0.5e^{-t} - 0.121306(1-r), 0.5e^{-t} + 0.121306(1-r)],$$
$$r \in [0,1], \quad t \in [0, 0.5].$$

The exact solution $x^* : [-0.5, 0.5] \to E$ *is given by*

$$x^*(t,r) = [\underline{x^*}(t,r), \overline{x^*}(t,r)] = [e^{-t} - 0.2(1-r), e^{-t} + 0.2(1-r)],$$
$$r \in [0,1], \quad t \in [-0.5, 0.5].$$

The above problem has been solved by the Bernoulli wavelet method for $M_1 = M_2 = 4$, $k_1 = k_2 = 2$ *and then the results have been compared with that obtained by B-spline wavelet method for* $m = 4$. *Here we take* $r = 0.25,\ 0.5,\ 0.75$ *and calculate the absolute errors as* $|e_r| = |x(t,r) - x^*(t,r)|$. *This comparison has been presented in Table 9.9 and Table 9.10. The overall computational times for* $\underline{x}(t,r)$ *and* $\overline{x}(t,r)$ *are 1.342 seconds and 1.249 seconds respectively by the Bernoulli wavelet method. On the other hand, the overall computational times for* $\underline{x}(t,r)$ *and* $\overline{x}(t,r)$ *are 478.173 seconds and 498.079 seconds respectively by the B-spline wavelet method. Figures 9.3 and 9.4 show the approximate solutions of* $\underline{x}(t,r)$ *and* $\overline{x}(t,r)$ *respectively, for* $r = 0, 0.25, 0.5, 0.75, 1$.

For this example, we have calculated the error bound in view of the error estimation theorem. Here, $M_H = \|e^{-(t-s)}\|_\infty \leq 2$ and since, there presents a linear term in the integrand part, we can take $L = 1$. We can calculate the lowest error bound for truncated error term as $\|E_{n_1,m_1,n_2,m_2}\|_\infty \leq 1.61953$. Also, the error in Gauss-Legendre quadrature for $n = 10$ can be calculated as

$$\|_1E_{GL}(t,\xi,r) -_2 E_{GL}(t,\xi,r)\|_\infty = \frac{2^{2n+1}(n!)^4}{(2n+1)[(2n)!]^3} \|V^{(2n)}(t,\xi,r) - U^{(2n)}(t,\xi,r)\|_\infty$$
$$\leq 1.20259 \times 10^{-24} \|V^{(2n)}(t,\xi,r) - U^{(2n)}(t,\xi,r)\|_\infty.$$

Hence,

$$\|R(t,r)\|_\infty \leq 3.23906 + 1.20259 \times 10^{-24} \|V^{(2n)}(t,\xi,r) - U^{(2n)}(t,\xi,r)\|_\infty.$$

TABLE 9.9: Comparison of numerical solutions for $\underline{x}(t, r)$ in Example 9.4.3

t	Bernoulli wavelet method			B-spline wavelet method		
	$\lvert e_{0.25}\rvert$	$\lvert e_{0.5}\rvert$	$\lvert e_{0.75}\rvert$	$\lvert e_{0.25}\rvert$	$\lvert e_{0.5}\rvert$	$\lvert e_{0.75}\rvert$
0.1	5.77E-6	5.74E-6	5.70E-6	4.34E-5	4.34E-5	4.34E-5
0.2	3.31E-6	3.35E-6	3.39E-6	4.64E-5	4.64E-5	4.64E-5
0.3	2.96E-6	3.01E-6	3.05E-6	2.58E-5	2.57E-5	2.57E-5
0.4	5.90E-6	5.85E-6	5.80E-6	1.80E-5	1.80E-5	1.81E-5
0.5	5.18E-5	5.19E-5	5.19E-5	7.40E-5	7.40E-5	7.41E-5

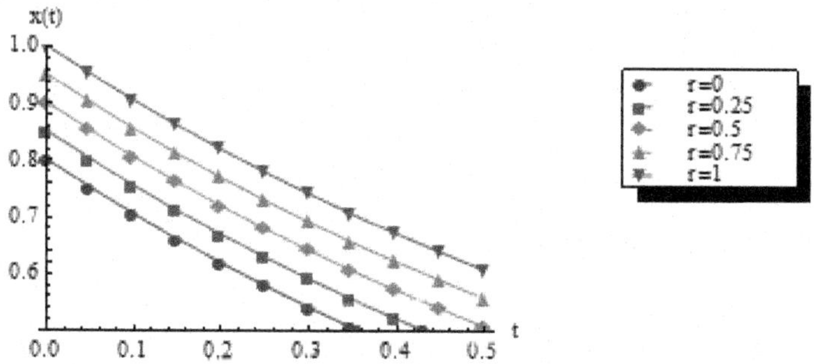

FIGURE 9.3: Approximate solution of $\underline{x}(t, r)$ for $r = 0, 0.25, 0.5, 0.75, 1$ of Example 9.4.3

9.5 Numerical solution of Fuzzy integro-differential equations

In this section, we consider the fuzzy integro-differential equation of the form

$$u'(t) + u(t) + \lambda \int_0^1 k(s,t)u(s)ds = g(t), \tag{9.39}$$
$$k(s,t) \in C([0,1] \times [0,1]), \quad t \in [0,1], \quad \lambda \in \mathbb{R},$$

TABLE 9.10: Comparison of numerical solutions for $\overline{x}(t, r)$ in Example 9.4.3

t	Bernoulli wavelet method			B-spline wavelet method		
	$\lvert e_{0.25}\rvert$	$\lvert e_{0.5}\rvert$	$\lvert e_{0.75}\rvert$	$\lvert e_{0.25}\rvert$	$\lvert e_{0.5}\rvert$	$\lvert e_{0.75}\rvert$
0.1	5.56E-6	5.59E-6	5.63E-6	4.32E-5	4.33E-5	4.33E-5
0.2	3.55E-6	3.51E-6	3.47E-6	4.62E-5	4.62E-5	4.63E-5
0.3	3.22E-6	3.17E-6	3.13E-6	2.55E-5	2.56E-5	2.56E-5
0.4	5.62E-6	5.66E-6	5.71E-6	1.83E-5	1.82E-5	1.82E-5
0.5	5.21E-5	5.21E-5	5.20E-5	7.43E-5	7.42E-5	7.42E-5

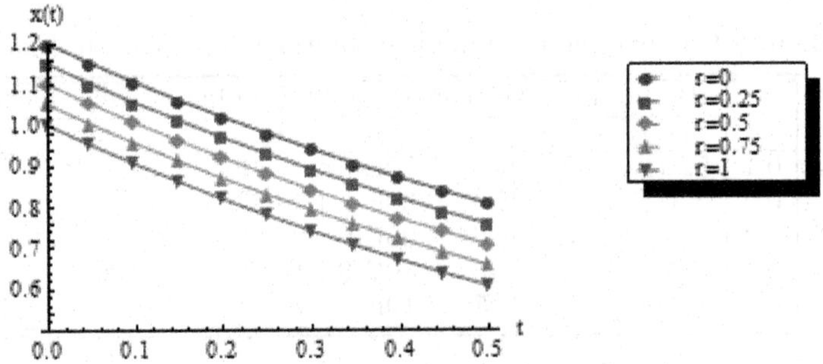

FIGURE 9.4: Approximate solution of $\overline{x}(t,r)$ for $r = 0, 0.25, 0.5, 0.75, 1$ of Example 9.4.3

with the initial condition $u(0) = \alpha$, where u and g are fuzzy functions, and the sign of $k(s,t)$ does not change in $[0,1]$. Let

$$u(t,r) = \left(\underline{u}(t,r), \overline{u}(t,r)\right),$$
$$g(t,r) = \left(\underline{g}(t,r), \overline{g}(t,r)\right),$$
$$\frac{\partial}{\partial t}u(t,r) = \left(\frac{\partial}{\partial t}\underline{u}(t,r), \frac{\partial}{\partial t}\overline{u}(t,r)\right),$$

where all derivatives with respect to t, be fuzzy functions.

Also the initial condition $u(0,r) = (\underline{u}(0,r), \overline{u}(0,r))$. Let

$$u^c(t,r) = \frac{\underline{u}(t,r) + \overline{u}(t,r)}{2},$$
$$u^d(t,r) = \frac{\overline{u}(t,r) - \underline{u}(t,r)}{2}, \tag{9.40}$$

and

$$g^c(t,r) = \frac{\underline{g}(t,r) + \overline{g}(t,r)}{2},$$
$$g^d(t,r) = \frac{\overline{g}(t,r) - \underline{g}(t,r)}{2}. \tag{9.41}$$

From eq. (9.39), the fuzzy integro-differential equation can be written as

$$\frac{\partial}{\partial t}u^c(t,r) + u^c(t,r) + \lambda \int_0^1 k(s,t)u^c(s,r)ds$$
$$= g^c(t,r),$$
$$\frac{\partial}{\partial t}u^d(t,r) + u^d(t,r) + \lambda \int_0^1 |k(s,t)|u^d(s,r)ds$$
$$= g^d(t,r), \tag{9.42}$$

with initial conditions

$$u^c(0,r) = \frac{\underline{u}(0,r) + \overline{u}(0,r)}{2},$$
$$u^d(0,r) = \frac{\overline{u}(0,r) - \underline{u}(0,r)}{2}. \quad (9.43)$$

9.5.1 Legendre wavelet method for fuzzy integro-differential equation

Consider the fuzzy integro-differential equation given in eq. (9.42). First, we integrate eq. (9.42) on both sides with respect to t from 0 to t and using the initial conditions from eq. (9.43), we have

$$u^c(t,r) - u^c(0,r) + \int_0^t u^c(t,r)dt$$
$$+ \lambda \int_0^t \int_0^1 k(s,t)u^c(s,r)dsdt = \int_0^t g^c(t,r)dt. \quad (9.44)$$

In order to apply the Legendre wavelets in eq. (9.44), we first approximate the unknown function $u^c(t,r)$ as

$$u^c(t,r) = C^T\Psi(t,r), \quad (9.45)$$

where C is defined similar to eq. (2.32) of Chapter 2.

Putting eq. (9.45) in eq. (9.44), we have

$$C^T\Psi(t,r) - u^c(0,r) + \int_0^t C^T\Psi(t,r)dt$$
$$+ \lambda \int_0^t \int_0^1 k(s,t)C^T\Psi(s,r)dsdt$$
$$= \int_0^t g^c(t,r)dt. \quad (9.46)$$

Now we collocate the eq. (9.46) at $(2^{k_1-1}2^{k_2-1}M_1M_2)$ points by $t_i = \frac{2i-1}{2^{k_1}M_1}$, $r_j = \frac{2j-1}{2^{k_2}M_2}$, for $i = 1,2,...,2^{k_1-1}M_1$, $j = 1,2,...,2^{k_2-1}M_2$ yielding

$$C^T\Psi(t_i,r_j) - u^c(0,r_j) + \int_0^{t_i} C^T\Psi(t,r_j)dt$$
$$+ \lambda \int_0^{t_i} \int_0^1 k(s,t)C^T\Psi(s,r_j)dsdt$$
$$= \int_0^{t_i} g^c(t,r_j)dt. \quad (9.47)$$

Eq. (9.47) constitutes a system of $\left(2^{k_1-1}2^{k_2-1}M_1M_2\right)$ algebraic equations with the same number of unknowns for coefficient matrix C. Solving this system numerically, we can get the values of unknowns for C and hence we obtain the solution $u^c(t,r) = C^T\Psi(t,r)$.

Similarly, we can obtain the solution of $u^d(t,r)$ for the second fuzzy integro-differential equation defined in eq. (9.42).

Hence we can obtain the solutions as fuzzy number

$$\underline{u}(t,r) = u^c(t,r) - u^d(t,r),$$
$$\overline{u}(t,r) = u^c(t,r) + u^d(t,r).$$

9.5.2 Convergence analysis

Theorem 9.5.1. *The series solution*

$$u(x) \cong \sum_{n=1}^{2^{k-1}} \sum_{m=0}^{M-1} c_{n,m}\psi_{n,m}(x)$$

using the Legendre wavelet method converges to $u(x)$

Proof. See Theorem 6.2.1 of Chapter 6. □

Theorem 9.5.2. *If $u(x,t)$ is a multiplicatively separable function, then the series solution*

$$u(x,t) \cong \sum_{n_1=0}^{2^{k_1-1}} \sum_{m_1=0}^{M_1-1} \sum_{n_2=0}^{2^{k_2-1}} \sum_{m_2=0}^{M_2-1} c_{n_1,m_1,n_2,m_2}\psi_{n_1,m_1,n_2,m_2}(x,t)$$

defined in eq. (2.34) of Chapter 2 using the two dimensional Legendre wavelet method converges to $u(x,t)$.

Proof. From eq. (2.29) of Chapter 2, the two dimensional Legendre wavelets can be expressed as

$$\psi_{n_1,m_1,n_2,m_2}(x,t) = \begin{cases} AP_{m_1}(2^{k_1}x - 2n_1 + 1)P_{m_2}(2^{k_2}t - 2n_2 + 1), \\ \quad \frac{n_1-1}{2^{k_1-1}} \le x < \frac{n_1}{2^{k_1-1}},\ \frac{n_2-1}{2^{k_2-1}} \le x < \frac{n_2}{2^{k_2-1}} \\ 0, \qquad \text{otherwise} \end{cases}$$
$$= \psi_{n_1,m_1}(x)\psi_{n_2,m_2}(t) \tag{9.48}$$

where $A = \sqrt{\left(m_1 + \frac{1}{2}\right)\left(m_2 + \frac{1}{2}\right)2^{\frac{k_1+k_2}{2}}}$.

Since c_{n_1,m_1,n_2,m_2} is arbitrary, we can decompose the two dimensional Legendre wavelets into a product of two one dimensional Legendre wavelets as

$$u(x,t) = \sum_{n_1=0}^{2^{k_1-1}} \sum_{m_1=0}^{M_1-1} \sum_{n_2=0}^{2^{k_2-1}} \sum_{m_2=0}^{M_2-1} c_{n_1,m_1,n_2,m_2} \psi_{n_1,m_1,n_2,m_2}(x,t)$$
$$= \left(\sum_{n_1=0}^{2^{k_1-1}} \sum_{m_1=0}^{M_1-1} d_{n_1,m_1} \psi_{n_1,m_1}(x) \right) \left(\sum_{n_2=0}^{2^{k_2-1}} \sum_{m_2=0}^{M_2-1} d'_{n_2,m_2} \psi_{n_2,m_2}(t) \right). \tag{9.49}$$

Since $u(x,t)$ is a multiplicatively separable function and from Theorem 9.5.1, the right hand side of Eq. (9.49) is convergent to the exact results. □

9.5.3 Illustrative examples

Example 9.5.1. *Consider the fuzzy integro-differential equation of the form*

$$\frac{\partial}{\partial t} u(t,r) + u(t,r) - \int_0^1 stu(s,r)ds = g(t,r), \qquad 0 < t, r \leq 1,$$

where

$$u(t,r) = (\underline{u}(t,r), \overline{u}(t,r)),$$
$$\underline{g}(t,r) = r + \frac{2rt}{3},$$
$$\overline{g}(t,r) = 8 - r - 4t - \frac{2rt}{3},$$

with initial conditions $\underline{u}(0,r) = 0$, *and* $\overline{u}(0,r) = 8$. *The exact solution of the above fuzzy integro-differential equation is given by*

$$u(t,r) = (\underline{u}(t,r), \overline{u}(t,r))$$
$$= (rt, 8 - rt).$$

From eqs. (9.40)-(9.41), *we have*

$$g^c(t,r) = \frac{\underline{g}(t,r) + \overline{g}(t,r)}{2}$$
$$= 4 - 2t,$$

$$g^d(t,r) = \frac{\overline{g}(t,r) - \underline{g}(t,r)}{2}$$
$$= 4 - r - 2t - \frac{2rt}{3},$$

and exact solutions of related crisp equations are

$$u^c(t,r) = \frac{\underline{u}(t,r) + \overline{u}(t,r)}{2} = 4,$$
$$u^d(t,r) = \frac{\overline{u}(t,r) - \underline{u}(t,r)}{2} = 4 - rt.$$

Here, initial conditions are $u^c(0,r) = 4$, $u^d(0,r) = 4$.
In the homotopy analysis method [2], the mth *order deformation equation approximating* $u^c(t,r)$ *is given by*

$$\begin{aligned} L\left[u_m^c(t,r) - \chi_m u_{m-1}^c(t,r)\right] &= \hbar \Re_m\left(u_0^c, u_1^c, ..., u_{m-1}^c\right) \\ &= \hbar \frac{1}{(m-1)!}\frac{\partial^{m-1}}{\partial q^{m-1}} N(\varphi(t,r;q))|_{q=0}, \end{aligned}$$

where

$$N(\varphi(t,r;q)) = \frac{\partial}{\partial t}\varphi(t,r;q) + \varphi(t,r;q) - \int_0^1 st\varphi(s,r;q)ds - 4 + 2t$$

and $\chi_m = \begin{cases} 0, & m \leq 1, \\ 1, & m > 1. \end{cases}$
Similarly, the mth *order deformation equation approximating* $u^d(t,r)$ *is given by*

$$\begin{aligned} L\left[u_m^d(t,r) - \chi_m u_{m-1}^d(t,r)\right] &= \hbar \Re_m\left(u_0^d, u_1^d, ..., u_{m-1}^d\right) \\ &= \hbar \frac{1}{(m-1)!}\frac{\partial^{m-1}}{\partial q^{m-1}} N(\varphi(t,r;q))|_{q=0}, \end{aligned}$$

where

$$\begin{aligned} N(\varphi(t,r;q)) = & \frac{\partial}{\partial t}\varphi(t,r;q) + \varphi(t,r;q) \\ & - \int_0^1 st\varphi(s,r;q)ds - 4 + r + 2t + \frac{2rt}{3}. \end{aligned}$$

Here auxiliary parameter [2] $\hbar = -1$ *belongs to the convergence region of the HAM series solution.*

The numerical results obtained by the Legendre wavelet method (LWM) for $M_1 = M_2 = 4$, $k_1 = k_2 = 1$ *have been compared with the results obtained by the* 6th *order homotopy analysis method (HAM), for* $r = 0.3,\ 0.5,\ 0.9$ *of* $u(t,r) = (\underline{u}(t,r), \overline{u}(t,r))$, $u^c(t,r)$, *and* $u^d(t,r)$. *These results have been shown in Table 9.11 and Table 9.13. Table 9.12 and Table 9.14 cite the absolute errors obtained by these two methods. Absolute error graphs for* $r = 0.3$ *of* $\underline{u}(t,r)$ *and* $r = 0.5$ *of* $\overline{u}(t,r)$ *have been shown in Figure 9.5 and Figure 9.6, respectively.*

TABLE 9.11: Numerical results of $u^c(t,r)$ and $u^d(t,r)$ obtained by LWM and HAM along with exact results for Example 9.5.1

r	t	$u^c(t,r)$			$u^d(t,r)$		
		Exact	HAM	LWM	Exact	HAM	LWM
0.3	0.2	4	4	4	3.94	3.94	3.94
	0.4	4	4	4	3.88	3.88	3.88
	0.6	4	4	4	3.82	3.82	3.82
	0.8	4	4	4	3.76	3.76	3.76
0.5	0.2	4	4	4	3.9	3.9	3.9
	0.4	4	4	4	3.8	3.8	3.8
	0.6	4	4	4	3.7	3.7	3.7
	0.8	4	4	4	3.6	3.6	3.6
0.9	0.2	4	4	4	3.82	3.82	3.82
	0.4	4	4	4	3.64	3.64	3.64
	0.6	4	4	4	3.46	3.46001	3.46
	0.8	4	4	4	3.28	3.27999	3.28

TABLE 9.12: Absolute errors of $u^c(t,r)$ and $u^d(t,r)$ obtained by LWM and HAM for Example 9.5.1

r	t	$u^c(t,r)$		$u^d(t,r)$	
		HAM	LWM	HAM	LWM
0.3	0.2	0	1.4635E-10	3.05754E-9	1.31835E-10
	0.4	0	4.81366E-10	7.10810E-9	4.20427E-10
	0.6	0	8.90564E-10	7.41406E-8	7.48768E-10
	0.8	0	1.27325E-9	2.05514E 7	1.01363E-9
0.5	0.2	0	1.46435E-10	6.17141E-7	1.22101E-10
	0.4	0	4.81366E-10	2.36169E-6	3.79800E-10
	0.6	0	8.90564E-10	2.85239E-6	6.54237E-10
	0.8	0	1.27325E-9	3.03736E-6	8.40549E-10
0.9	0.2	0	1.46435E-10	1.11085E-6	1.02634E-10
	0.4	0	4.81366E-10	4.25104E-6	2.98547E-10
	0.6	0	8.90564E-10	5.13431E-6	4.65173E-10
	0.8	0	1.27325E-9	5.46725E-6	4.94384E-10

TABLE 9.13: Numerical results of $u(t,r) = (\ \underline{u}(t,\ r),\ \overline{u}(t,r)\ )$ obtained by LWM and HAM along with exact results for Example 9.5.1

r	t	$\underline{u}(t,r)$			$\overline{u}(t,r)$		
		Exact	HAM	LWM	Exact	HAM	LWM
0.3	0.2	0.06	0.0599996	0.06	7.94	7.94	7.94
	0.4	0.12	0.119999	0.12	7.88	7.88	7.88
	0.6	0.18	0.179998	0.18	7.82	7.82	7.82
	0.8	0.24	0.240002	0.24	7.76	7.76	7.76
0.5	0.2	0.1	0.0999994	0.1	7.9	7.9	7.9
	0.4	0.2	0.199998	0.2	7.8	7.8	7.8
	0.6	0.3	0.299997	0.3	7.7	7.7	7.7
	0.8	0.4	0.400003	0.4	7.6	7.6	7.6
0.9	0.2	0.18	0.179999	0.18	7.82	7.82	7.82
	0.4	0.36	0.359996	0.36	7.64	7.64	7.64
	0.6	0.54	0.539995	0.54	7.46	7.46001	7.46
	0.8	0.72	0.72005	0.72	7.28	7.27999	7.28

TABLE 9.14: Absolute errors of $u(t,r) = (\ \underline{u}(t,\ r),\ \overline{u}(t,r)\ )$ obtained by LWM and HAM for Example 9.5.1

r	t	$\underline{u}(t,r)$		$\overline{u}(t,r)$	
		HAM	LWM	HAM	LWM
0.3	0.2	3.70285e-7	1.46003E-11	3.70285E-7	2.78268E-10
	0.4	1.41701E-6	6.09398E-11	1.41701E-6	9.01792E-10
	0.6	1.71144E-6	1.41797E-10	1.71144E-6	1.63933E-9
	0.8	1.82242E-6	2.59624E-10	1.82242E-6	2.28688E-9
0.5	0.2	6.17141E-7	2.43338E-11	6.17141E-7	2.68535E-10
	0.4	2.36169E-6	1.01566E-10	2.36169E-6	8.61165E-10
	0.6	2.85239E-6	2.36329E-10	2.85239E-6	1.5448E-9
	0.8	3.03736E-6	4.32706E-10	3.03736E-6	2.1138E-9
0.9	0.2	1.11085E-6	4.38007E-11	1.11085E-6	2.49067E-10
	0.4	4.25104E-6	1.82819E-10	4.25104E-6	7.79912E-10
	0.6	5.13431E-6	4.25391E-10	5.13431E-6	1.35574E-9
	0.8	5.46725E-6	7.78871E-10	5.46725E-6	1.76764E-9

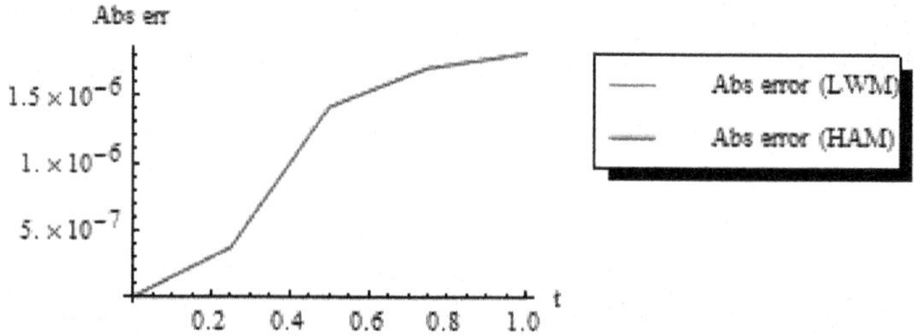

FIGURE 9.5: Absolute error graphs for $\underline{u}(t,\, r)$, $r = 0.3$ for Example 9.5.1.

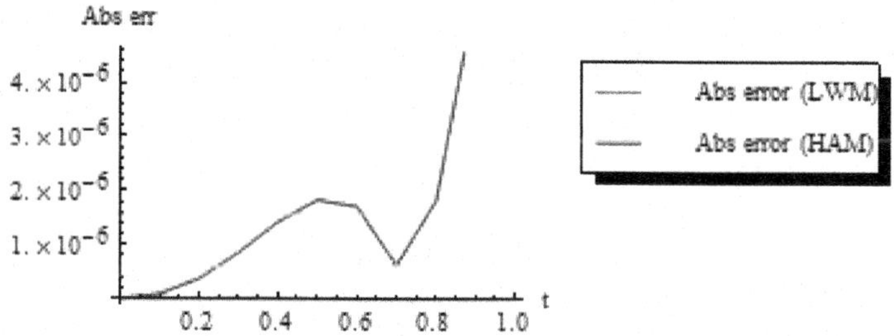

FIGURE 9.6: Absolute error graphs for $\overline{u}(t, r)$, $\quad r = 0.5$ for Example 9.5.1.

Example 9.5.2. *Consider the fuzzy integro-differential equation of the form*

$$\frac{\partial}{\partial t}u(t,r) + u(t,r) + \int_0^1 s^2 t u(s,r)ds = g(t,r), \quad 0 \le t, r \le 1,$$

where

$$\begin{aligned} u(t,r) &= \left(\underline{u}(t,r), \overline{u}(t,r)\right), \\ \underline{g}(t,r) &= r + \frac{5rt}{4}, \\ \overline{g}(t,r) &= 12 - 5r + 4t - \frac{25rt}{4}, \end{aligned}$$

with initial conditions $\underline{u}(0,r) = 0$, *and* $\overline{u}(0,r) = 12$. *The exact solution of the above fuzzy integro-differential equation is given by*

$$\begin{aligned} u(t,r) &= (\underline{u}(t,r), \overline{u}(t,r)) \\ &= (rt, 12 - 5rt) \end{aligned}$$

From eqs. (9.40)-(9.41), *we have*

$$g^c(t,r) = \frac{\underline{g}(t,r) + \overline{g}(t,r)}{2}$$
$$= 6 - 2r + 2t - \frac{5rt}{2},$$

$$g^d(t,r) = \frac{\overline{g}(t,r) - \underline{g}(t,r)}{2}$$
$$= 6 - 3r + 2t - \frac{15rt}{4},$$

and exact solutions of related crisp equations are

$$u^c(t,r) = \frac{\underline{u}(t,r) + \overline{u}(t,r)}{2} = 6 - 2rt,$$
$$u^d(t,r) = \frac{\overline{u}(t,r) - \underline{u}(t,r)}{2} = 6 - 3rt.$$

Here, initial conditions are $u^c(0,r) = 6$, $u^d(0,r) = 6$.
In the homotopy analysis method [2], the mth *order deformation equation approximating* $u^c(t,r)$ *is given by*

$$L\left[u^c_m(t,r) - \chi_m u^c_{m-1}(t,r)\right] = \hbar\Re_m\left(u^c_0, u^c_1, ..., u^c_{m-1}\right)$$
$$= \hbar\frac{1}{(m-1)!}\frac{\partial^{m-1}}{\partial q^{m-1}}N(\varphi(t,r;q))|_{q=0},$$

where

$$N(\varphi(t,r;q)) = \frac{\partial}{\partial t}\varphi(t,r;q) + \varphi(t,r;q)$$
$$+ \int_0^1 s^2 t\varphi(s,r;q)ds - 6 - 2r + 2t - \frac{5rt}{2},$$

and $\chi_m = \begin{cases} 0, & m \leq 1, \\ 1, & m > 1. \end{cases}$

Similarly, the mth *order deformation equation approximating* $u^d(t,r)$ *is given by*

$$L\left[u^d_m(t,r) - \chi_m u^d_{m-1}(t,r)\right] = \hbar\Re_m\left(u^d_0, u^d_1, ..., u^d_{m-1}\right)$$
$$= \hbar\frac{1}{(m-1)!}\frac{\partial^{m-1}}{\partial q^{m-1}}N(\varphi(t,r;q))|_{q=0},$$

where

$$N(\varphi(t,r;q)) = \frac{\partial}{\partial t}\varphi(t,r;q) + \varphi(t,r;q)$$
$$+ \int_0^1 s^2 t\varphi(s,r;q)ds - 6 + 3r + 2t - \frac{15rt}{4}.$$

TABLE 9.15: Numerical results of $u^c(t,r)$ and $u^d(t,r)$ obtained by LWM and HAM along with exact results for Example 9.5.2

r	t	$u^c(t,r)$			$u^d(t,r)$		
		Exact	HAM	LWM	Exact	HAM	LWM
0.3	0.2	5.88	5.88002	5.88	5.82	5.82	5.82
	0.4	5.76	5.76012	5.76	5.64	5.64002	5.64
	0.6	5.64	5.64035	5.64	5.46	5.46005	5.46
	0.8	5.52	5.52084	5.52	5.28	5.28012	5.28
0.5	0.2	5.8	5.80004	5.8	5.7	5.70001	5.7
	0.4	5.6	5.6002	5.6	5.4	5.40003	5.4
	0.6	5.4	5.40058	5.4	5.1	5.10008	5.1
	0.8	5.2	5.20140	5.2	4.8	4.8002	4.8
0.9	0.2	5.64	5.64007	5.64	5.46	5.46001	5.46
	0.4	5.28	5.28035	5.28	4.92	4.92005	4.92
	0.6	4.92	4.92105	4.92	4.38	4.38015	4.38
	0.8	4.56	4.56253	4.56	3.84	3.84037	3.84

Here auxiliary parameter [2] $\hbar = -1$ *belongs to the convergence region of the HAM series solution.*
The numerical results obtained by the Legendre wavelet method (LWM) for $M_1 = M_2 = 4$, $k_1 = k_2 = 1$ *have been compared with the results obtained by the* 6th *order homotopy analysis method (HAM), for* $r = 0.3,\ 0.5,\ 0.9$ *of* $u(t,r) = (\underline{u}(t,r), \overline{u}(t,r))$, $u^c(t,r)$, *and* $u^d(t,r)$. *These results have been shown in Table 9.15 and Table 9.17. Table 9.16 and Table 9.18 cite the absolute errors obtained by these two methods. Absolute error graphs for* $r = 0.3$ *of* $\underline{u}(t,r)$ *and* $r = 0.5$ *of* $\overline{u}(t,r)$ *have been shown in Figure 9.7 and Figure. 9.8, respectively.*

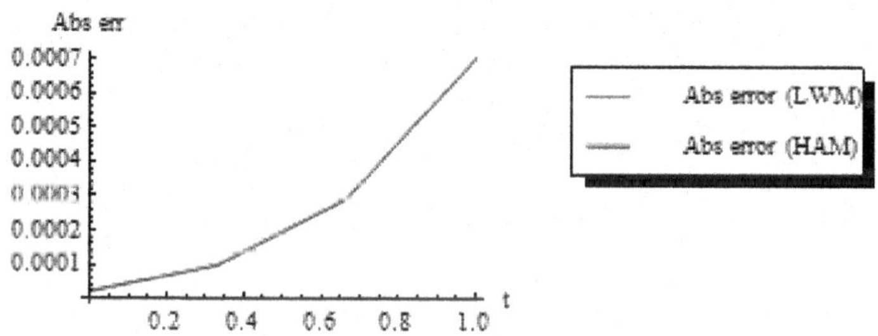

FIGURE 9.7: Absolute error graphs for $\underline{u}(t,\ r)$, $r = 0.3$ for Example 9.5.2.

TABLE 9.16: Absolute errors of $u^c(t,r)$ and $u^d(t,r)$ obtained by LWM and HAM for Example 9.5.2

r	t	$u^c(t,r)$		$u^d(t,r)$	
		HAM	LWM	HAM	LWM
0.3	0.2	2.30142E-5	1.94623E-10	3.45213E-5	1.82434E-10
	0.4	1.18062E-4	7.66436E-10	1.77092E-4	7.08815E-10
	0.6	3.49652E-4	1.70225E-9	5.24479E-4	1.55626E-9
	0.8	8.41839E-4	2.99043E-9	1.26276E-3	2.70457E-9
0.5	0.2	3.8357E-5	1.78370E-10	5.75355E-5	1.58054E-10
	0.4	1.96769E-4	6.89608E-10	2.95154E-4	5.93574E-10
	0.6	5.82754E-4	1.50759E-9	8.74131E-4	1.26427E-9
	0.8	1.40307E-3	2.60928E-9	2.1046E-3	2.13285E-9
0.9	0.2	6.90426E-5	1.45862E-10	1.03564E-4	1.09294E-10
	0.4	3.54185E-4	5.35951E-10	5.31277E-4	3.63089E-10
	0.6	1.04896E-3	1.11828E-9	1.57344E-3	6.80306E-10
	0.8	2.52552E-3	1.84699E-9	3.78828E-3	9.89412E-10

TABLE 9.17: Numerical results of $u(t,r) = (\ \underline{u}(t,\ r),\ \overline{u}(t,r)\ )$ obtained by LWM and HAM along with exact results for Example 9.5.2

r	t	$\underline{u}(t,r)$			$\overline{u}(t,r)$		
		Exact	HAM	LWM	Exact	HAM	LWM
0.3	0.2	0.06	0.599885	0.06	11.7	11.7001	11.7
	0.4	0.12	0.119941	0.12	11.4	11.4003	11.4
	0.6	0.18	0.179825	0.18	11.1	11.1009	11.1
	0.8	0.24	0.239579	0.24	10.8	10.8021	10.8
0.5	0.2	0.1	0.0999808	0.1	11.5	11.5001	11.5
	0.4	0.2	0.199902	0.2	11.0	11.0005	11.0
	0.6	0.3	0.299709	0.3	10.5	10.5015	10.5
	0.8	0.4	0.399298	0.4	10.0	10.0035	10.0
0.9	0.2	0.18	0.179965	0.18	11.1	11.1002	11.1
	0.4	0.36	0.359823	0.36	10.2	10.2009	10.2
	0.6	0.54	0.539476	0.54	9.3	9.30262	9.3
	0.8	0.72	0.718737	0.72	8.4	8.40631	8.4

TABLE 9.18: Absolute errors of $u(t,r) = (\ \underline{u}(t,\ r),\ \overline{u}(t,r)\ )$ obtained by LWM and HAM for Example 9.5.2

r	t	$\underline{u}(t,r)$		$\overline{u}(t,r)$	
		HAM	LWM	HAM	LWM
0.3	0.2	1.15071E-5	1.21899E-11	5.75355E-5	3.77055E-10
	0.4	5.90308E-5	5.76211E-11	2.95154E-4	1.47525E-9
	0.6	1.74826E-4	1.45992E-10	8.74131E-4	3.2585E-9
	0.8	4.2092E-4	2.85586E-10	2.1046E-3	5.695E-9
0.5	0.2	1.91785E-5	2.03164E-11	9.58925E-5	3.36422E-10
	0.4	9.83847E-5	9.60351E-11	4.91923E-4	1.28318E-9
	0.6	2.91377E-4	2.43321E-10	1.45688E-3	2.77186E-9
	0.8	7.01533E-4	4.76434E-10	3.50766E-3	4.74214E-9
0.9	0.2	3.45213E-5	3.65696E-11	1.72606E-4	2.55156E-10
	0.4	1.77092E-4	1.72863E-10	8.85462E-4	8.99036E-10
	0.6	5.24479E-4	4.37977E-10	2.62239E-3	1.79858E-9
	0.8	1.26276E-3	8.5758E-10	6.3138E-3	2.8364E-9

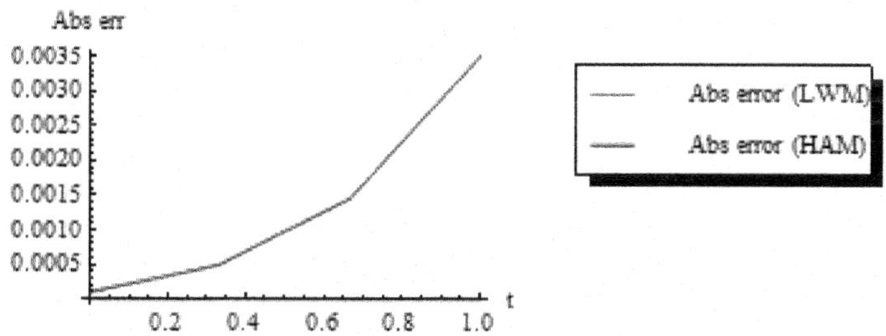

FIGURE 9.8: Absolute error graphs for $\overline{u}(t,r)$ $r = 0.5$ for Example 9.5.2.

9.6 Conclusion

In this chapter, many numerical techniques based on polynomial approximation and orthogonal wavelets approximation have been implemented to solve fuzzy integral equations and fuzzy intgro-differential equations. In Section 9.3, the Bernstein polynomial collocation method and the Legendre wavelet method have been applied to the nonlinear Hammerstein fuzzy Fredholm integral equations and the obtained results then compared with the results obtained by the homotopy analysis method. The presented methods reduce the Hammerstein fuzzy Fredholm integral equation to a system of nonlinear algebraic equations and this system has been solved by Newton's method. Since the homotopy analysis method is an analytical method, then from the tables, it is verified that the results obtained by presented methods are very accurate with regard to HAM results and also it is cleared that the Bernstein polynomial collocation method gives more accuracy than the Legendre wavelet method. In Section 9.4, the Bernoulli wavelet method has been applied to solve nonlinear fuzzy Hammerstein-Volterra delay integral equations. This method reduces the integral equations to a system of algebraic equations and that algebraic system has been solved by Newton's method. Also the numerical results obtained by the present method have been compared with the results obtained by the B-spline wavelet method. Although the similar procedure has been implemented using both the methods to solve these delay integral equations, from the results, it manifests that the present Bernoulli wavelet method gives more accurate results than B-spline wavelet method results. Additionally, the computational time of the present method is significantly less than that obtained by the B-spline wavelet method. Moreover, we can get very less absolute error by increasing the order of the Bernoulli polynomials. In Section 9.5, the two dimensional Legendre wavelet method has been applied to solve the fuzzy integro-differential equations. Using this procedure the integral equations have been reduced to solve a system of algebraic equations. From the tables and figures, it is clear that the obtained numerical results by the present method highly agree with exact solutions and are more efficient than the HAM solutions. We can get more accuracy by increasing the value of M_1, M_2, k_1, k_2. The illustrative examples have been included to demonstrate the validity and applicability of the proposed technique. These examples also exhibit the accuracy and efficiency of the two dimensional Legendre wavelet method.

Chapter 10

Numerical Solutions of Fractional Integro-Differential Equations

10.1 Introduction

Fractional calculus is an emerging and popular field among the science and engineering community. Many physical problems [156, 47, 44, 194] are modeled by fractional differential equations and fractional integral equations, and obtaining solutions of these equations have been the subject of many researchers in recent years. Frequently many researchers search lot of dynamical problems that exhibit fractional order behavior varying with time and space. It has been applied to model the nonlinear oscillation of earthquakes, fluid-dynamic traffic, frequency dependent damping behavior of many viscoelastic materials, continuum and statistical mechanics, solid mechanics, economics, signal processing, and control theory [26, 121, 165, 27, 111]. These fractional integral equations and integro-differential equations have been solved both analytically and numerically [74, 174, 177]. Many analytical methods like the Adomian decomposition method [141], the homotopy perturbation method [189], the homotopy analysis method [82], the variational iteration method [104], the fractional differential transform method (FDTM) [14], the generalized block pulse operational matrix method [16], and the Laplace transform method [156] have been developed to solve fractional integral equations and integro-differential equations. But the analytical solutions of fractional integral or integro-differential equations are not obtainable always. That is the main reason why finding the numerical solutions of these problems has become a great deal for many researchers.

Many numerical techniques have been developed to solve fractional integral equations and integro-differential equations. Nonlinear fractional Fredholm integro-differential equations [231] and nonlinear fractional Volterra integro-differential equations [230] with initial conditions have been solved by using the second kind Chebyshev wavelet. The Legendre wavelet method [83] has been applied to solve a fractional population growth model which is in the form of fractional Fredholm integro-differential equations. Fractional integro-differential equations are also solved by the cubic B-spline wavelet method [135]. The Petrov-Galerkin method is an efficient method that has been applied by many researchers to solve different types of partial differential equations, integral equations and integro-differential equations. An hp Petrov-Galerkin finite element method [214] has been applied to solve Volterra integro-differential equations. In [112], linear Volterra integro-differential equations have been solved by the Petrov-Galerkin method. Also, the Petrov-Galerkin method has been successively applied to solve many partial differential equations (see Refs. [227, 164, 163, 85, 117]). Previously, the Sinc-Galerkin method has been applied to solve nonlinear boundary value problems by Gamel et al. [59]. The Sinc-Galerkin method has been applied to fourth order differential equations by Bowers et al. in [195]. Recently, Secer et al. [190] have employed the Sinc-Galerkin technique to fractional order boundary value problems successively. The Sinc collocation method has been applied to multi-order fractional differential equations in [81]. Solving space-fractional boundary value problems, the Sinc-Galerkin method has been developed by Alkan and Secer [9]. There are several studies about the application of the Sinc function based method which can be found in [168, 159, 143, 80].

In this chapter, we have solved fractional integro-differential equations of different types by different numerical methods. In Section 10.2, the preliminaries of fractional calculus have been discussed. In Section 10.3, we have considered the nonlinear fractional mixed Volterra-Fredholm integro-differential equations along with mixed boundary conditions and solved them by the Legendre wavelet method. These types of fractional Volterra-Fredholm integro-differential equations with mixed boundary conditions have been solved by the Nystrom method [146] for a large value of n. In this section, we have compared the numerical results obtained by the Legendre wavelet method and the Nystrom method [146]. In Section 10.4, we have developed a numerical scheme using the Petrov-Galerkin method where the trial and test functions are Legendre wavelets basis functions. Also, this method has been applied to solve fractional Volterra integro-differential equations. The uniqueness and existence of the problem have been discussed and the error estimate of the proposed method has been presented in this section. In Section 10.5, the Sinc-Galerkin method is developed to approximate the solution of fractional Volterra-Fredholm integro-differential equations with weakly singular kernels. The proposed method is based on the Sinc function approximation. The numerical results obtained by the Sinc-Galerkin method have been compared with the results obtained by existing methods. The uniqueness and existence of the problem have been discussed and the error analysis of the proposed method has been presented in this section. Some illustrative examples have been provided to show the applicability and accuracy of the present method. In Section 10.6, concluding remarks have been discussed.

10.2 Preliminaries of fractional calculus

In this section, some important definitions of fractional derivatives and integrations have been discussed. The fractional calculus involves different definitions of the fractional operators as well as the Riemann-Liouville fractional derivative, the Caputo derivative, the Riesz derivative and the Grunwald-Letnikov fractional derivative [140, 188, 157]. The fractional calculus has gained considerable importance during the past decades mainly due to its applications in diverse fields of science and engineering. In this chapter, the Caputo's definition of fractional derivative has been used, considering the advantage of Caputo's approach that the initial conditions for fractional differential equations with Caputo's derivatives take on the traditional form as that for integer order differential equations.

Definition 10.2.1. (Riemann-Liouville fractional integral) *The Riemann-Liouville fractional integral [157] of order* $\alpha > 0$ *of a function* f *is defined* *as*

$$J^{\alpha} f(t) = \frac{1}{\Gamma(\alpha)} \int_0^t (t-\tau)^{\alpha-1} f(\tau) d\tau, \quad t > 0, \quad \alpha \in \mathbb{R}^+ \tag{10.1}$$

where $\mathbb{R}^+$ *is the set of positive real numbers.*

Definition 10.2.2. (Riemann-Liouville fractional derivative) *The Riemann-Liouville fractional derivative of order* $\alpha > 0$ *is normally defined as*

$$D^{\alpha} f(t) = D^m J^{m-\alpha} f(t), \quad m-1 < \alpha \leq m, \tag{10.2}$$

where $m \in \mathbb{N}$.

Definition 10.2.3. (Caputo fractional derivative) *The fractional derivative, introduced by Caputo [48, 46] in the late sixties, is called the Caputo fractional derivative. The fractional derivative of* $f(t)$ *in the Caputo sense is defined by*

$$\begin{aligned} D_t^{\alpha} f(t) &= J^{m-\alpha} D^m f(t) \\ &= \begin{cases} \frac{1}{\Gamma(m-\alpha)} \int_0^t (t-\tau)^{m-\alpha-1} \frac{d^m f(\tau)}{d\tau^m} d\tau, & m-1 < \alpha < m, \ m \in \mathbb{N}, \\ \frac{d^m f(t)}{dt^m}, & \alpha = m, \ m \in \mathbb{N}. \end{cases} \end{aligned} \tag{10.3}$$

where the parameter α *is the order of the derivative and is allowed to be real or even complex. In this work, only real and positive* α *will be considered.*

For the Caputo derivative we have

$$D^{\alpha} C = 0, \qquad (C \text{ is a constant})$$

$$D^{\alpha} t^{\beta} = \begin{cases} 0, & \beta \leq \alpha - 1, \\ \frac{\Gamma(\beta+1) t^{\beta-\alpha}}{\Gamma(\beta-\alpha+1)}, & \beta > \alpha - 1. \end{cases}$$

Similar to integer order differentiation, the Caputo derivative is linear.

$$D^{\alpha}(\gamma f(t) + \delta g(t)) = \gamma D^{\alpha} f(t) + \delta D^{\alpha} g(t),$$

where γ and δ are constants, and satisfies the so called Leibnitz rule.

$$D^{\alpha}(g(t) f(t)) = \sum_{k=0}^{\infty} \binom{\alpha}{k} g^{(k)}(t) D^{\alpha-k} f(t),$$

if $f(\tau)$ is continuous in $[0,t]$ and $g(\tau)$ has sufficient number of continuous derivatives in $[0,t]$.

Lemma 10.2.1. [173] *Let* $Re(\alpha) > 0$ *and let* $n = [Re(\alpha)] + 1$ *for* $\alpha \notin N_0 = \{0, 1, 2, ...\}$; $n = \alpha$ *for* $\alpha \in N_0$. *If* $f(t) \in AC^n[a,b]$ *(the space of functions* $f(t)$ *which are absolutely continuous and possess continuous derivatives up to* *order $n-1$ on $[a,b]$) or $f(t) \in C^n[a,b]$ (the space of functions $f(t)$ which are n times continuously differentiable on $[a,b]$), then*

$$^{C}D_t^{\alpha} J^{\alpha} f(t) = f(t) \tag{10.4}$$

and

$$J^{\alpha}\, {}^{C}D_t^{\alpha} f(t) = f(t) - \sum_{k=0}^{n-1} \frac{t^k}{k!} f^{(k)}(0+), \quad t > 0. \tag{10.5}$$

10.3 Numerical solutions of nonlinear fractional Volterra-Fredholm integro-differential equations with mixed boundary conditions

In this section, we have considered the following form of nonlinear fractional mixed Volterra-Fredholm integro-differential equation and solved by the Legendre wavelet method.

$$(D^{\alpha}y)(x) = g(x) + \int_0^x k_1(x,t)F_1[t,y(t)]dt + \int_0^1 k_2(x,t)F_2[t,y(t)]dt \tag{10.6}$$

with mixed boundary conditions

$$\sum_{j=1}^{d}[a_{i,j}y^{(j-1)}(0) + b_{i,j}y^{(j-1)}(1)] = r_i, \quad i = 1,2,...,d, \tag{10.7}$$

where $y : [0,1] \to \mathbb{R}$ is the continuous function which has to be determined, $g : [0,1] \to \mathbb{R}$ and $k_i : [0,1] \times [0,1] \to \mathbb{R}$, $i = 1,2$ are continuous functions. $F_i : [0,1] \times \mathbb{R} \to \mathbb{R}$, $i = 1,2$ are nonlinear terms and Lipschitz continuous functions. Here D^{α} is understood as the Caputo fractional derivative. Using Legendre wavelets this fractional integro-differential equation is converted into an algebraic equations system which again can be solved by Newton's method. The properties of Legendre wavelets and their function approximation have been discussed in Chapter 2. Note that for easy simplification, we have applied the Gauss Legendre quadrature rule for evaluating the integration on nonlinear terms. The obtained results again compared with that of the Nystrom method.

10.3.1 Existence and uniqueness

Consider the fractional Volterra integro-differential equation (10.6) and rewrite the eq. (10.6) in operator form as

$$D^{\alpha}y(x) = g(x) + \mathcal{K}_1\mathcal{F}_1 y + \mathcal{K}_2\mathcal{F}_2 y, \tag{10.8}$$

where

$$\mathcal{K}_1\mathcal{F}_1 y = \int_0^x K_1(x,t)F_1[y(t)]dt,$$

$$\mathcal{K}_2\mathcal{F}_2 y = \int_0^1 K_2(x,t)F_2[y(t)]dt.$$

Applying J^{α} on both sides of eq. (10.8), we have

$$y(x) = h(x) + J^{\alpha}\left[g(x) + \mathcal{K}_1\mathcal{F}_1 y + \mathcal{K}_2\mathcal{F}_2 y\right], \tag{10.9}$$

where $h(x) = \sum_{k=0}^{n-1} \frac{t^k}{k!} y^{(k)}(0+)$, $n-1 < \alpha < n$, $n \in \mathbb{N}$. Eq. (10.9) can be written as fixed point equation form $\mathcal{A}y = y$, where $\mathcal{A}$ is defined as

$$\mathcal{A}y(x) = h(x) + J^{\alpha}\left[g(x) + \mathcal{K}_1\mathcal{F}_1 y + \mathcal{K}_2\mathcal{F}_2 y\right]. \tag{10.10}$$

Let $(C[0,1], \|.\|_{\infty})$ be the Banach space of all continuous functions with norm $\|f\|_{\infty} = \max_t |f(t)|$. Also, the operators $\mathcal{F}_1$ and $\mathcal{F}_2$ satisfy the Lipschitz condition on $[0,1]$ as

$$|\mathcal{F}_1\tilde{y}_m(x) - \mathcal{F}_1 y(x)| \le L_1|\tilde{y}_m(x) - y(x)|,$$

$$|\mathcal{F}_2\tilde{y}_m(x) - \mathcal{F}_2 y(x)| \le L_2|\tilde{y}_m(x) - y(x)|,$$

where L_1 and L_2 are Lipschitz constants. Then we have proceed to prove the uniqueness of the solution of the eq. (10.6).

Theorem 10.3.1. *If* $L_1\|\mathcal{K}_1\|_{\infty} + L_2\|\mathcal{K}_2\|_{\infty} < \Gamma(\alpha+1)$, *then the problem* (10.6) *has an unique solution* $y \in [0,1]$.

Proof. Let $\mathcal{A} : C[0,1] \to C[0,1]$ such that

$$\mathcal{A}y(x) = h(x) + \frac{1}{\Gamma(\alpha)}\int_0^x (x-t)^{\alpha-1}\left[g(t) + \mathcal{K}_1\mathcal{F}_1 y(t) + \mathcal{K}_2\mathcal{F}_2 y(t)\right]dt.$$

Let $\tilde{y}, y \in C[0,1]$ and

$$\mathcal{A}\tilde{y}(x) - \mathcal{A}y(x) = \frac{1}{\Gamma(\alpha)}\int_0^x (x-t)^{\alpha-1} \times \left[\left[\mathcal{K}_1\mathcal{F}_1\tilde{y}(t) - \mathcal{K}_1\mathcal{F}_1 y(t)\right] + \left[\mathcal{K}_2\mathcal{F}_2\tilde{y}(t) - \mathcal{K}_2\mathcal{F}_2 y(t)\right]\right]dt.$$

Then for $x > 0$, we have

$$
\begin{aligned}
|\mathcal{A}\tilde{y}(x) - \mathcal{A}y(x)| &\leq \frac{1}{\Gamma(\alpha)} \int_0^x |x-t|^{\alpha-1} \\
&\quad \times [|\mathcal{K}_1||\mathcal{F}_1\tilde{y}(t) - \mathcal{F}_1 y(t)| + |\mathcal{K}_2||\mathcal{F}_2\tilde{y}(t) - \mathcal{F}_2 y(t)|]\, dt \\
&\leq \frac{1}{\Gamma(\alpha)} \int_0^x |x-t|^{\alpha-1} \\
&\quad \times [|\mathcal{K}_1|L_1|\tilde{y}(t) - y(t)| + |\mathcal{K}_2|L_2|\tilde{y}(t) - y(t)|]\, dt \\
&\leq \frac{1}{\Gamma(\alpha)} \int_0^x |x-t|^{\alpha-1}(L_1\|\mathcal{K}_1\|_\infty + L_2\|\mathcal{K}_2\|_\infty)\|\tilde{y} - y\|_\infty \\
&\leq (L_1\|\mathcal{K}_1\|_\infty + L_2\|\mathcal{K}_2\|_\infty)\|\tilde{y} - y\|_\infty \frac{|x|^\alpha}{\Gamma(\alpha+1)} \\
&\leq (L_1\|\mathcal{K}_1\|_\infty + L_2\|\mathcal{K}_2\|_\infty)\|\tilde{y} - y\|_\infty \frac{1}{\Gamma(\alpha+1)}.
\end{aligned}
$$

Therefore,

$$\|\mathcal{A}\tilde{y}(x) - \mathcal{A}y(x)\|_\infty \leq \Omega_{L_1,L_2,\mathcal{K}_1,\mathcal{K}_2,\alpha}\|\tilde{y} - y\|_\infty,$$

where

$$\Omega_{L_1,L_2,\mathcal{K}_1,\mathcal{K}_2,\alpha} = (L_1\|\mathcal{K}_1\|_\infty + L_2\|\mathcal{K}_2\|_\infty)\frac{1}{\Gamma(\alpha+1)}.$$

Since $\Omega_{L_1,L_2,\mathcal{K}_1,\mathcal{K}_2,\alpha} < 1$, by the contraction mapping theorem, the problem (10.6) has an unique solution in $C[0,1]$. □

10.3.2 Legendre wavelet method for fractional Volterra-Fredholm integro-differential equation

Let us consider the nonlinear fractional Volterra-Fredholm integro-differential equation with mixed boundary conditions given in equations (10.6)-(10.7) and we approximate the unknown function $y(x) \in [0,1]$ by the Legendre wavelet method as

$$y(x) = C^T\Psi(x). \tag{10.11}$$

We assume

$$F_1[y(x)] = u(x), \tag{10.12}$$

$$F_2[y(x)] = v(x). \tag{10.13}$$

Again we approximate $u(x)$ and $v(x)$ by (2.31) of Chapter 2 as

$$u(x) = A_1^T\Psi(x), \tag{10.14}$$

$$v(x) = A_2^T\Psi(x), \tag{10.15}$$

where A_1 and A_2 are similar to C defined in eq. (2.32) of Chapter 2.

First, applying the J^α both sides of the eq. (10.6) and using the eqs. (10.11)-(10.15), we have

$$(J^\alpha D^\alpha y)(x) = J^\alpha[g(x)] + J^\alpha[\int_0^x k_1(x,t)F_1[y(t)]dt] + J^\alpha[\int_0^1 k_2(x,t)F_2[y(t)]dt]$$

$$y(x) - \sum_{l=0}^{d-1} \frac{x^l}{l!} y^{(l)}(0+) = \frac{1}{\Gamma(\alpha)} \int_0^x (x-\tau)^{\alpha-1} g(\tau)d\tau + \frac{1}{\Gamma(\alpha)} \int_0^x (x-\tau)^{\alpha-1} \int_0^\tau k_1(\tau,t)u(t)dtd\tau + \frac{1}{\Gamma(\alpha)} \int_0^x (x-\tau)^{\alpha-1} \int_0^1 k_2(\tau,t)v(t)dtd\tau \quad (10.16)$$

The exact solution of eq. (10.16) has been replaced with the approximate solution $C^T\Psi(x)$ as

$$C^T\Psi(x) - \sum_{l=0}^{d-1} \frac{x^l}{l!} C^T\Psi^{(l)}(0+) = \frac{1}{\Gamma(\alpha)} \int_0^x (x-\tau)^{\alpha-1} g(\tau)d\tau + \frac{1}{\Gamma(\alpha)} \int_0^x (x-\tau)^{\alpha-1} \int_0^\tau k_1(\tau,t)A_1^T\Psi(t)dtd\tau + \frac{1}{\Gamma(\alpha)} \int_0^x (x-\tau)^{\alpha-1} \int_0^1 k_2(\tau,t)A_2^T\Psi(t)dtd\tau. \quad (10.17)$$

Putting collocation points $x_i = \frac{2i-1}{2^k M}$, $i = 1, 2, ..., 2^{k-1}M$ in eq. (10.17), we have

$$C^T\Psi(x_i) - \sum_{l=0}^{d-1} \frac{x_i^l}{l!} C^T\Psi^{(l)}(0+) = \frac{1}{\Gamma(\alpha)} \int_0^{x_i} (x_i-\tau)^{\alpha-1} g(\tau)d\tau + \frac{1}{\Gamma(\alpha)} \int_0^{x_i} (x_i-\tau)^{\alpha-1} \int_0^\tau k_1(\tau,t)A_1^T\Psi(t)dtd\tau + \frac{1}{\Gamma(\alpha)} \int_0^{x_i} (x_i-\tau)^{\alpha-1} \int_0^1 k_2(\tau,t)A_2^T\Psi(t)dtd\tau \quad (10.18)$$

Now, to apply the Gauss-Legendre quadrature rule for evaluating the integrals in eq. (10.18), we change the domain of integration from $[0, x_i]$ to $[-1, 1]$ using the transformation $\tau = \frac{x_i}{2}(s+1)$ and then apply the Gauss-Legendre rule yielding

$$C^T\Psi(x_i) - \sum_{l=0}^{d-1} \frac{x_i^l}{l!} C^T\Psi^{(l)}(0+) = \frac{1}{\Gamma(\alpha)} \int_0^{x_i} (x_i - \tau)^{\alpha-1} g(\tau) d\tau$$
$$+ \frac{1}{\Gamma(\alpha)} \frac{x_i}{2} \sum_{j=1}^{M_1} w_j \left(\frac{x_i}{2}(1-s_j)\right)^{\alpha-1} \int_0^{\frac{x_i}{2}(s_j+1)} k_1\left(\frac{x_i}{2}(s_j+1), t\right) A_1^T \Psi(t) dt$$
$$+ \frac{1}{\Gamma(\alpha)} \frac{x_i}{2} \sum_{j=1}^{M_2} w_j \left(\frac{x_i}{2}(1-s_j)\right)^{\alpha-1} \int_0^{1} k_2\left(\frac{x_i}{2}(s_j+1), t\right) A_2^T \Psi(t) dt, \tag{10.19}$$
$$i = 1, 2, ..., 2^{k-1}M,$$

where M_1 and M_2 are the order of Legendre polynomial used in the Gauss-Legendre quadrature rule.

Next, from eqs. (10.12)-(10.13), we have

$$F_1[C^T\Psi(x)] = A_1^T\Psi(x),$$
$$F_2[C^T\Psi(x)] = A_2^T\Psi(x). \tag{10.20}$$

Using the eqs. (10.20) by collocation points $x_i = \frac{2i-1}{2^k M}$, $i = 1, 2, ..., 2^{k-1}M$, we get

$$F_1[C^T\Psi(x_i)] = A_1^T\Psi(x_i),$$
$$F_2[C^T\Psi(x_i)] = A_2^T\Psi(x_i). \tag{10.21}$$

Again from boundary conditions, we have

$$\sum_{j=1}^{d} [a_{i,j} y^{(j-1)}(0+) + b_{i,j} y^{(j-1)}(1-)] = r_i, \quad i = 1, 2, ..., d \tag{10.22}$$

Combining eqs. (10.19)-(10.22) gives a system of $3 \times 2^{k-1}M$ number of nonlinear algebraic equations with the same number of unknowns in the vectors C, A_1 and A_2. Numerically solving this system by Newton's method, we get the solutions for the unknown vectors C, A_1 and A_2. Hence we obtain the approximate solution $y(x)$ from eq. (10.11).

10.3.3 Convergence analysis

Theorem 10.3.2. *Let $y(x)$ be a function defined on $[0,1)$ and $|y(x)| \le \mathcal{M}_y$; then the sum of absolute value of Legendre coefficients of $y(x)$ defined in eq. (2.31) of Chapter 2 converges absolutely on the interval $[0,1]$ if*

$$|c_{n,m}| \le 2^{\frac{1-k}{2}} \mathcal{M}_y.$$

Proof. Any function $y(x) \in L^2[0,1]$ can be approximated by Legendre wavelets as

$$y(x) \cong \sum_{n=1}^{2^{k-1}} \sum_{m=0}^{M-1} c_{n,m}\psi_{n,m}(x),$$

where the coefficients $c_{n,m}$ can be determined as

$$c_{n,m} = \langle y(x), \psi_{n,m}(x)\rangle.$$

Now for $m \geq 0$,

$$\begin{aligned}
|c_{n,m}| &= |\langle y(x), \psi_{n,m}(x)\rangle| \\
&= |\int_0^1 y(x)\psi_{n,m}(x)dx| \\
&\leq \int_0^1 |y(x)||\psi_{n,m}(x)|dx \\
&\leq \mathcal{M}_y \int_0^1 |\psi_{n,m}(x)|dx \\
&= \mathcal{M}_y \int_{I_{nk}} |\psi_{n,m}(x)|dx \\
&= \mathcal{M}_y \sqrt{m+\frac{1}{2}}\ 2^{k/2} \int_{I_{nk}} |P_m(2^k x - 2n + 1)|dx,
\end{aligned}$$

where $I_{nk} = [\frac{n-1}{2^{k-1}}, \frac{n}{2^{k-1}})$.

Now, changing the variable $2^k x - 2n + 1 = t$, we have

$$|c_{n,m}| \leq \mathcal{M}_y \sqrt{m+\frac{1}{2}}\ 2^{-k/2} \int_{-1}^1 |P_m(t)|dt.$$

By applying Hölder's inequality,

$$\begin{aligned}
\left(\int_{-1}^1 |P_m(t)|dt\right)^2 &\leq \left(\int_{-1}^1 1^2 dt\right)\left(\int_{-1}^1 |P_m(t)|^2 dt\right) \\
&= 2 \times \frac{2}{2m+1} \\
&= \frac{4}{2m+1}.
\end{aligned}$$

This implies that

$$\int_{-1}^1 |P_m(t)|dt \leq \frac{2}{\sqrt{2m+1}}.$$

Hence,

$$|c_{n,m}| \leq 2^{\frac{1-k}{2}} \mathcal{M}_y.$$

This means that the series $\sum_{n=1}^{2^{k-1}} \sum_{m=0}^{M-1} c_{n,m}$ is absolutely convergent as $k \to \infty$. □

Theorem 10.3.3. *If the sum of the absolute values of the Legendre coefficients of a continuous function $y(x)$ forms a convergent series, then the Legendre expansion $\sum_{n=1}^{2^{k-1}} \sum_{m=0}^{M-1} c_{n,m}\psi_{n,m}(x)$ converges with respect to L^2-norm on $[0,1]$.*

Proof. Let $L^2(\mathbb{R})$ be the Hilbert space and $\psi_{n,m}$ defined in (2.28) of Chapter 2 forms an orthonormal basis.
Let

$$\tilde{y}(x) = \sum_{n=1}^{2^{k-1}} \sum_{m=0}^{M-1} c_{n,m}\psi_{n,m}(x),$$

where $c_{n,m} = \langle \tilde{y}(x), \psi_{n,m}(x)\rangle$ for a fixed n.

Let us denote $\psi_{n,m}(x) = \chi_j(x)$ and let $\alpha_j = \langle \tilde{y}(x), \chi_j(x)\rangle$.
Now we define the sequence of partial sums $\{S_n\}$, where

$$S_n(x) = \sum_{j=0}^{n} \alpha_j \chi_j(x).$$

For every $\epsilon > 0$ there exists a positive number $N(\epsilon)$ such that for every $n > m > N(\epsilon)$,

$$\begin{aligned} \|S_n(x) - S_m(x)\|_2^2 &= \int_0^1 |\sum_{k=m+1}^{n} \alpha_k \chi_k(x)|^2 dx \\ &\leq \sum_{k=m+1}^{n} |\alpha_k|^2 \int_0^1 |\chi_k(x)|^2 dx \\ &= \sum_{k=m+1}^{n} |\alpha_k|^2. \end{aligned}$$

From Theorem 10.3.2, $\sum_{k=0}^{\infty} |\alpha_k|^2$ is absolutely convergent.
According to the Cauchy criterion, for every $\epsilon > 0$, there exists a positive number $N(\epsilon)$ such that

$$\sum_{k=m+1}^{n} |\alpha_k|^2 < \epsilon,$$

whenever $n > m > N(\epsilon)$.

Hence,

$$\|S_n(x) - S_m(x)\|_2^2 \leq \sum_{k=m+1}^{n} |\alpha_k|^2 < \epsilon$$

TABLE 10.1: Comparison of absolute errors for Example 10.3.1

x	Error by LWM for $M = 8, k = 2$	Error by [146] for $n = 320$
0.1	1.33492E-6	1.92E-6
0.3	2.28941E-6	3.84E-6
0.5	6.23726E-8	4.10E-6
0.7	1.81063E-6	3.15E-6
0.9	6.8395E-6	1.25E-6

This implies that

$$\|S_n(x) - S_m(x)\|_2 \leq \sqrt{\epsilon} < \epsilon$$

Thus, the sequence of partial sum of the series converges with respect to L^2-norm and hence it completes the proof. □

10.3.4 Illustrative examples

In this section, we have considered three test problems from which Example 10.3.1 is taken from the literature (see ref. [146]) that verifies the accuracy of the present method.

Example 10.3.1. *Let us consider the nonlinear fractional Volterra-Fredholm integro-differential equation*

$$(D^{\frac{\sqrt{7}}{2}}y)(x) = g(x) + \int_0^x \frac{1+2t}{1+y(t)}dt + \int_0^1 (1+2t)e^{y(t)}dt,$$

with boundary conditions $y(0) = 0$, $y(1) = 2$, *and*

$$g(x) = 1 - e^2 - \log(1 + x + x^2) - \frac{4x^{2+\frac{\sqrt{7}}{2}}}{(\sqrt{7}-4)\Gamma(2-\frac{\sqrt{7}}{2})}.$$

The exact solution of this problem is given as $y(x) = x + x^2$. *This problem has been solved by the Legendre wavelet method (LWM) which reduces the integral equation to a system of algebraic equations that has been solved by the Newton method. Again the results obtained by the present method have been compared with that of the Nystrom method [146]. The numerical results and absolute errors for Example 10.3.1 have been presented in Table 10.1.*

Example 10.3.2. *Let us consider the nonlinear fractional Volterra-Fredholm integro-differential equation*

$$(D^{1.7}y)(x) = g(x) + \int_0^x x^2 t \log[y(t)]dt + \int_0^1 [x + 2ty^2(t)]dt,$$

TABLE 10.2: Comparison of numerical results and absolute errors for Example 10.3.2

x	Exact	LWM ($M=4, k=2$)		Nystrom method($N=20$)	
		$y(x)$	Abs. Error	$y(x)$	Abs. Error
0.0	1	1	0	1	0
0.1	1.10517	1.10506	0.00011483	1.10393	0.00123938
0.2	1.2214	1.22073	0.00066919	1.21842	0.00297924
0.3	1.34986	1.34835	0.00151145	1.34476	0.00510241
0.4	1.49182	1.48921	0.00261231	1.4843	0.00752611
0.5	1.64872	1.64255	0.00617023	1.63862	0.0101061
0.6	1.82212	1.81694	0.00518308	1.80958	0.0125403
0.7	2.01375	2.00622	0.00752974	1.99952	0.0142327
0.8	2.22554	2.21589	0.00964998	2.21144	0.0141018
0.9	2.4596	2.45142	0.00818527	2.44928	0.0103248

with boundary conditions $y(0) = 1$, $y(1) = e$ *and*

$$g(x) = -4.19453 - x - \frac{1}{3}x^5 + \frac{1}{\Gamma(1.3)} e^x x^{0.3}\ {}_1F_1[0.3, 1.3; -x],$$

where ${}_1F_1[0.3, 1.3; -x]$ *is the Kummer confluent hypergeometric function and defined as*

$$ {}_1F_1[a, b; z] = \sum_{n=0}^{\infty} \frac{(a)_n}{(b)_n} \frac{z^n}{n!},$$

with $(a)_n = a(a+1)(a+2)...(a+n-1)$ *and* $(a)_0 = 1$.

The exact solution of this problem is given as $y(x) = e^x$. *This problem has been solved by the Legendre wavelet method (LWM) for* $M = 4$, $k = 2$, *which reduces the integral equation to a system of algebraic equations that has been solved by the Newton method. Again the results obtained by the present method have been compared with that of the Nystrom method (for* $N = 20$*). The numerical results and absolute errors for Example 10.3.2 have been presented in Table 10.2.*

Example 10.3.3. *As the third example, we consider the nonlinear fractional Volterra-Fredholm integro-differential equation*

$$(D^{\sqrt{3}}y)(x) = -\frac{x^2}{6} - \frac{15x^8}{56} + \frac{2(2+\sqrt{3})x^{2-\sqrt{3}}}{\Gamma(2-\sqrt{3})} + \int_0^x (x+t)y^3(t)dt + \int_0^1 x^2 t y^2(t)dt,$$

subject to the mixed boundary conditions

$$y(0) + y'(0) = 0,$$
$$y(1) + y'(1) = 3,$$

TABLE 10.3: Comparison of numerical results and absolute errors for Example 10.3.3

x	Exact	LWM ($M = 4, k = 2$)		Nystrom method($N = 20$)	
		$y(x)$	Abs. Error	$y(x)$	Abs. Error
0.0	0	0.000766339	0.000766339	0	0
0.1	0.01	0.0106897	0.000689691	0.0100005	4.87E-7
0.2	0.04	0.040613	0.000613028	0.0400065	6.50E-6
0.3	0.09	0.0905364	0.000536369	0.0900304	0.0000304128
0.4	0.16	0.16046	0.000459735	0.160101	0.00010141
0.5	0.25	0.250372	0.000371695	0.250329	0.000329115
0.6	0.36	0.360307	0.000306562	0.361156	0.00115625
0.7	0.49	0.490222	0.000221624	0.494128	0.00412784
0.8	0.64	0.640136	0.000136065	0.653945	0.0139455
0.9	0.81	0.810069	0.000069068	0.85402	0.0440196

with the exact solution $y(x) = x^2$. This problem has been solved by the Legendre wavelet method (LWM) for $M = 4$, $k = 2$, which reduces the integral equation to a system of algebraic equations that has been solved by the Newton method. Again the results obtained by the present method have been compared with that of the Nystrom method (for $N = 20$). The numerical results and absolute errors for Example 10.3.3 have been presented in Table 10.3.

10.4 Legendre wavelet Petrov-Galerkin method for fractional Volterra integro-differential equations

In this section, we have developed the Petrov-Galerkin method where the trial and test functions are Legendre wavelets basis functions. Also, this method has been applied to solve fractional Volterra integro-differential equations. These types of integro-differential equations are very difficult to solve analytically. The authors have considered the following form of fractional Volterra integro-differential equation

$$(D_x^\alpha y)(x) = g(x) + p(x)y(x) + \int_0^x K(x,t)y(t)dt, \quad y(0) = y_0, \qquad (10.23)$$

where $y(x)$ is the unknown function and $p(x)$, $g(x)$ are known functions. Here D_x^α is understood as the Caputo fractional derivative. Using the Legendre wavelets Petrov-Galerkin method (LWPGM), this fractional integro-differential equation is converted into a system of algebraic equations which again can be solved by Newton's method. Note that for easy implementation, we have applied the Gauss-Legendre quadrature rule for evaluating the integrations.

10.4.1 Existence and uniqueness

Consider the fractional Volterra integro-differential equation (10.23) and we rewrite the eq. (10.23) in operator form as

$$D_x^\alpha y(x) = g(x) + p(x)\mathcal{F}y + \mathcal{K}y, \tag{10.24}$$

where

$$\mathcal{K}y = \int_0^x K(x,t)y(t)dt.$$

Applying J^α on both sides of eq. (10.24), we have

$$y(x) = h(x) + J^\alpha \left[g(x) + p(x)\mathcal{F}y + \mathcal{K}y\right], \tag{10.25}$$

where $h(x) = \sum_{k=0}^{n-1} \frac{t^k}{k!} y^{(k)}(0+)$. Eq. (10.25) can be written as fixed point equation form $\mathcal{A}y = y$, where $\mathcal{A}$ is defined as

$$\mathcal{A}y(x) = h(x) + J^\alpha \left[g(x) + p(x)\mathcal{F}y + \mathcal{K}y\right]. \tag{10.26}$$

Let $(C[0,1], \|.\|_\infty)$ be the Hilbert space of all continuous functions with norm $\|f\|_\infty = \max_t |f(t)|$. Also, the operator $\mathcal{F}$ satisfies the Lipschitz condition on $[0,1]$ as

$$|\mathcal{F}\tilde{y}_m(x) - \mathcal{F}y(x)| \leq L|\tilde{y}_m(x) - y(x)|,$$

where L is Lipschitz constant. Then we proceed to prove the uniqueness of the solution of the eq. (10.23).

Theorem 10.4.1. *If* $L\|p\|_\infty + \|K\|_\infty < \Gamma(\alpha+1)$, *then the initial value problem* (10.23) *has an unique solution* $y \in [0,1]$.

Proof. Let $\mathcal{A} : C[0,1] \to C[0,1]$ such that

$$\mathcal{A}y(x) = h(x) + \frac{1}{\Gamma(\alpha)} \int_0^x (x-t)^{\alpha-1} \left[g(t) + p(t)\mathcal{F}y(t) + \mathcal{K}y(t)\right] dt.$$

Let $\tilde{y}, y \in C[0,1]$ and

$$\mathcal{A}\tilde{y}(x) - \mathcal{A}y(x) = \frac{1}{\Gamma(\alpha)} \int_0^x (x-t)^{\alpha-1} \times \left[p(t)[\mathcal{F}\tilde{y}(t) - \mathcal{F}y(t)] + \mathcal{K}[\tilde{y}(t) - y(t)]\right] dt.$$

Then for $x > 0$, we have

$$\begin{aligned}|\mathcal{A}\tilde{y}(x) - \mathcal{A}y(x)| &\leq \frac{1}{\Gamma(\alpha)}\int_0^x |x-t|^{\alpha-1} \\ &\quad \times [|p(t)|L|\tilde{y}(t) - y(t)| + |\mathcal{K}||\tilde{y}(t) - y(t)|]\, dt \\ &\leq \frac{1}{\Gamma(\alpha)}\int_0^x |x-t|^{\alpha-1}(L\|p\|_\infty + \|\mathcal{K}\|_\infty)\|\tilde{y} - y\|_\infty \\ &\leq (L\|p\|_\infty + \|\mathcal{K}\|_\infty)\|\tilde{y} - y\|_\infty \frac{|x|^\alpha}{\Gamma(\alpha+1)} \\ &\leq (L\|p\|_\infty + \|\mathcal{K}\|_\infty)\|\tilde{y} - y\|_\infty \frac{1}{\Gamma(\alpha+1)}.\end{aligned}$$

Therefore,

$$\|\mathcal{A}\tilde{y}(x) - \mathcal{A}y(x)\|_\infty \leq \Omega_{L,p,\mathcal{K},\alpha}\|\tilde{y} - y\|_\infty,$$

where

$$\Omega_{L,p,\mathcal{K},\alpha} = (L\|p\|_\infty + \|\mathcal{K}\|_\infty)\frac{1}{\Gamma(\alpha+1)}.$$

Since $\Omega_{L,p,\mathcal{K},\alpha} < 1$, by the contraction mapping theorem, the initial value problem (10.23) has an unique solution in $C[0,1]$. □

10.4.2 Legendre wavelets Petrov-Galerkin method

Let

$$S_m^{(0)}[0,1] = \{v \in H^1[0,1] : v|_{[0,1]} \in \mathbf{P}_m\},$$
$$S_{m-1}^{(-1)}[0,1] = \{v \in L^2[0,1] : v|_{[0,1]} \in \mathbf{P}_{m-1}\}$$

be the trial and test function spaces, respectively, where $\mathbf{P}_m$ denotes the space of polynomials of degree not exceeding m. The superscript (-1) in the test function space emphasizes that it is not a subspace of $C[0,1]$. The LWPG solution of eq. (10.23) is defined on the function $y \in S_m^{(0)}[0,1]$, then $y' \in S_{m-1}^{(-1)}[0,1]$. Here $H^1[0,1]$ is the Sobolev space. Eq. (10.23) can be revolutionized by LWPGM as

$$\langle D_x^\alpha y, \varphi\rangle = \langle g, \varphi\rangle + \langle p(x)y, \varphi\rangle + \langle \int_0^x K(x,t)y(t)dt, \varphi\rangle, \tag{10.27}$$

where $y \in S_m^{(0)}[0,1]$ and $\varphi \in S_{m-1}^{(-1)}$ and $\langle . \rangle$ is the usual inner product in $L^2[0,1]$.

Now we approximate the unknown function $y(x)$ by Legendre wavelets defined in eq. (2.31) of Chapter 2 where n is fixed as 1.

$$\langle C^T[D_x^\alpha \Psi(x)], \varphi\rangle = \langle g(x), \varphi\rangle + \langle p(x)C^T\Psi(x), \varphi\rangle + \langle \int_0^x K(x,t)C^T\Psi(t)dt, \varphi\rangle, \tag{10.28}$$

where, $\varphi(x)$ can be taken as Legendre wavelets $\psi_{n,m}(x)$, $n = 1$, $m = 0, 1, ..., M-2$.

Eq. (10.28) gives $M-1$ equations with M unknowns. Again from the initial conditions we have

$$C^T\Psi(0) = 0. \tag{10.29}$$

From eqs. (10.28) and (10.29), we obtain a system of M algebraic equations with M unknowns. Solving this system by any numerical method, we obtain the approximate value of unknowns and hence determine the solution; we apply the Gauss-Legendre quadrature rule to obtain the integration in eq. (10.28).

10.4.3 Error estimate

Lemma 10.4.2. *For a continuous function $y : [0,1] \to \mathbb{R}$, there exists a piecewise polynomial $\mathcal{I}y \in S_m^{(0)}[0,1]$ such that $\mathcal{I}y(t) = y(t)$.*

Let us consider an auxiliary problem and find $\tilde{Y} \in S_m^{(0)}[0,1]$, such that

$$\int_I \tilde{Y}'(t)\varphi(t)dt = \int_I g(t)\varphi(t)dt + \int_I p(t)y(t)\varphi(t)dt + \int_I \mathcal{K}y(t)\varphi(t)dt, \tag{10.30}$$
$$\tilde{Y}(0) = y_0,$$

for all $\varphi \in S_{m-1}^{(-1)}(I)$. Let y be the exact solution and Y be the LWPGM solution of eq. (10.23). Let us split the error of the LWPGM into two parts, i.e.,

$$y - Y = \eta + \xi,$$

where $\eta = y - \tilde{Y}$ and $\xi = \tilde{Y} - Y$. Our aim to find the error bound for both η and ξ separately.

Error bound for η:

From eq. (10.23) and eq. (10.30), it holds

$$\int_I \eta'\varphi dt = 0, \qquad \text{for all} \quad \varphi \in S_{m-1}^{(-1)}[0,1]. \tag{10.31}$$

From eq. (10.31), we have

$$\int_I (\tilde{Y} - \mathcal{I}y)'\varphi dt = \int_I (y - \mathcal{I}y)'\varphi dt. \tag{10.32}$$

Choosing $\varphi = (\tilde{Y} - \mathcal{I}y)'$ yields $|\tilde{Y} - \mathcal{I}y|_{H^1[0,1]} \leq |y - \mathcal{I}y|_{H^1[0,1]}$, thus we have

$$\begin{aligned}|\eta|_{H^1[0,1]} = |y - \tilde{Y}|_{H^1[0,1]} &\leq |y - \mathcal{I}y|_{H^1[0,1]} + |\mathcal{I}y - \tilde{Y}|_{H^1[0,1]} \\ &\leq 2|y - \mathcal{I}y|_{H^1[0,1]}.\end{aligned}$$

Setting $\varphi = 1$, we have $\eta(t) - \eta(0) = 0$. Integrating by parts eq. (10.31), we have

$$\int_I \eta\varphi' dt = 0, \qquad \forall\varphi \in S_{m-1}^{(-1)}[0,1]. \tag{10.33}$$

For $m \geq 1$, we define a function ψ by

$$\psi(t) = \int_I \eta(\tau)d\tau,$$

then there exists a polynomial $\tilde{\psi} \in S_{m-1}^{(-1)}[0,1]$ such that

$$|\psi - \tilde{\psi}|_{H^1[0,1]} \leq C\frac{1}{m}|\psi|_{H^2[0,1]}. \tag{10.34}$$

Then from eq. (10.33), we obtain

$$\begin{aligned}\|\eta\|^2_{L^2[0,1]} &= \int_I \eta(\eta - \tilde{\psi}')dt \\ &= \int_I \eta(\psi' - \tilde{\psi}')dt \\ &\leq \|\eta\|_{L^2[0,1]}\|\psi' - \tilde{\psi}'\|_{L^2[0,1]}.\end{aligned}$$

This implies

$$\|\eta\|_{L^2[0,1]} \leq \|\psi' - \tilde{\psi}'\|_{L^2[0,1]}. \tag{10.35}$$

From eqs. (10.34) and (10.35), we have

$$\|\eta\|_{L^2[0,1]} \leq C\frac{1}{m}|\psi|_{H^2[0,1]} = C\frac{1}{m}|\eta|_{H^1[0,1]}.$$

Error bound for ξ:

For any $u \in L^2[0,1]$, we denote $\prod_{m-1} u \in S_{m-1}^{(-1)}[0,1]$, the L^2 projection of u onto $S_{m-1}^{(-1)}[0,1]$,

$$\int_I (u - \prod_{m-1} u)\varphi dt = 0, \qquad \forall\varphi \in S_{m-1}^{(-1)}[0,1].$$

Setting $\varphi = \prod_{m-1} u$ and using the Cauchy-Schwartz inequality, we obtain the L^2-stability of the projection operator $\prod_{m-1}$,

$$\|\prod_{m-1} u\|_{L^2[0,1]} \leq \|u\|_{L^2[0,1]}, \qquad \forall u \in L^2[0,1].$$

From the eqs. (10.23) and (2.31) of Chapter 2, there holds

$$\int_I \xi'\varphi dt = \int_I p(t)(y - Y)\varphi dt + \int_I \mathcal{K}(y - Y)\varphi dt, \quad \varphi \in S_{m-1}^{(-1)}[0,1].$$

Then choosing $\varphi = \prod_{m-1} \xi$ leads to

$$\begin{aligned}
\| \int_I \xi' \xi dt \| &= \| \int_I p(t)(y - Y) \prod_{m-1} \xi dt + \int_I \mathcal{K}(y - Y) \prod_{m-1} \xi dt \| \\
&\leq \|p\| \int_I |y - Y| | \prod_{m-1} \xi | dt + \|\mathcal{K}\| \int_I \int_0^t |y - Y| ds | \prod_{m-1} \xi | dt \\
&\leq \|p\| \|y - Y\|_{L^2[0,1]} \|\xi\|_{L^2[0,1]} + \|\mathcal{K}\| \|y - Y\|_{L^2[0,1]} \|\xi\|_{L^2[0,1]} \\
&= (\|p\| + \|\mathcal{K}\|) \|y - Y\|_{L^2[0,1]} \|\xi\|_{L^2[0,1]}.
\end{aligned}$$

This implies

$$\frac{1}{2}(\|\xi(t)\|^2 - \|\xi(0)\|^2) \leq (\|p\| + \|\mathcal{K}\|) \|y - Y\|_{L^2[0,1]} \|\xi\|_{L^2[0,1]}.$$

Hence

$$\|\xi\| \leq 2(\|p\| + \|\mathcal{K}\|) \|y - Y\|_{L^2[0,1]}.$$

10.4.4 Illustrative examples

Example 10.4.1. *Let us consider the fractional Volterra integro-differential equations [88]*

$$D_x^{\frac{1}{2}} y(x) = y(x) + \frac{8}{3\Gamma(0.5)} x^{1.5} - x^2 - \frac{1}{3} x^3 + \int_0^x y(t) dt,$$

with initial condition $y(0) = 0$. The exact solution of this problem is x^2. This problem has been solved by the Legendre wavelet Petrov-Galerkin method (for $M = 6$) which reduces the integral equation to a system of algebraic equations. The numerical results and absolute errors of Example 10.4.1 have been provided in Table 10.4.

Example 10.4.2. *Let us consider the fractional Volterra integro-differential equation [88]*

$$D_x^{0.75} y(x) = \frac{1}{\Gamma(1.25)} x^{0.25} + (x \cos x - \sin x) y(x) + \int_0^x x \sin t y(t) dt,$$

with initial condition $y(0) = 0$. the exact solution of this problem is x. This problem has been solved by the Legendre wavelet Petrov-Galerkin method (for $M = 6$) which reduces the integral equation to a system of algebraic equations. The numerical results and absolute errors of Example 10.4.2 have been provided in Table 10.5.

TABLE 10.4: Absolute errors for Example 10.4.1

x	Exact	Error by LWPGM
0.0	0.0	3.88578E-16
0.1	0.01	5.55112E-16
0.2	0.04	6.66134E-16
0.3	0.09	9.15934E-16
0.4	0.16	1.27676E-15
0.5	0.25	1.63758E-15
0.6	0.36	2.04003E-15
0.7	0.49	2.52576E-15
0.8	0.64	3.27516E-15
0.9	0.81	3.77476E-15
1.0	1.0	4.21885E-15

TABLE 10.5: Absolute errors for Example 10.4.2

x	Exact	Error by LWPGM
0.0	0.0	0.0
0.1	0.1	5.55112E-17
0.2	0.2	0.0
0.3	0.3	2.77556E-17
0.4	0.4	4.16334E-17
0.5	0.5	5.55112E-17
0.6	0.6	6.93889E-17
0.7	0.7	8.32667E-17
0.8	0.8	1.11022E-16
0.9	0.9	1.11022E-16
1.0	1.0	0.0

TABLE 10.6: Absolute errors for Example 10.4.3

x	Exact	Error by LWPGM
0.0	0.0	5.55112E-17
0.1	0.01	0
0.2	0.04	2.22045E-16
0.3	0.09	2.77556E-16
0.4	0.16	2.77556E-16
0.5	0.25	3.19189E-16
0.6	0.36	3.46945E-16
0.7	0.49	3.33067E-16
0.8	0.64	1.11022E-16
0.9	0.81	1.66533E-16
1.0	1.0	1.11022E-16

Example 10.4.3. *As the third example, let us consider the nonlinear fractional Volterra-Fredholm integro-differential equation [146]*

$$D_x^{\sqrt{3}} y(x) = -\frac{2}{\Gamma(3-\sqrt{3})} x^{2-\sqrt{3}} + 2\sin x - 2x + \int_0^x \cos(x-t) y(t) dt,$$

with initial condition $y(0) = 0$*. The exact solution of this problem is* x^2*. This problem has been solved by the Legendre wavelet Petrov-Galerkin method (for* $M = 6$*) which reduces the integral equation to a system of algebraic equations. The numerical results and absolute errors of Example 10.4.3 have been provided in Table 10.6.*

10.5 Sinc Galerkin technique for the numerical solution of fractional Volterra-Fredholm integro-differential equations with weakly singular kernels

In this section, we have employed the Sinc-Galerkin method for solving the Fredholm-Volterra integro-differential equation with weakly singular kernel which has the following form

$$D_t^{\alpha} u(t) = \lambda \int_0^t K_1(t,s) u(s) ds + \mu \int_0^1 K_2(t,s) u(s) ds + f(t), \quad 0 < \alpha < 1, \tag{10.36}$$

with $u(0) = u_0$ as the initial condition. Here, $u(t)$ is the unknown function, $K_2(t,s), f(t)$ are known continuous functions, $K_1(t,s)$ is a singular kernel function, and λ, μ are reals. In this case, D_t^α is understood as a Caputo fractional derivative. Using the Sinc-Galerkin method (SGM), this fractional integro-differential equation is converted into a system of algebraic equations which again can be solved by Newton's method.

10.5.1 Existence and uniqueness

Consider the fractional Volterra integro-differential equation (10.36) and we rewrite the eq. (10.36) in operator form as

$$D_t^\alpha u(t) = \lambda \mathcal{K}_1 u(t) + \mu \mathcal{K}_2 u(t) + \mathcal{F}(t), \tag{10.37}$$

where $u, f : (0,1) \to \mathbb{R}$ and $K_2 : (0,1) \times (0,1) \to \mathbb{R}$ are analytic functions. $K_1(t,s)$ is a singular kernel function which has to be shown as an analytic function.
Let $K_1(t,s) = g(t,s)|t-s|^{-\alpha}$ where g is a Lipschitz continuous function, Then

$$\begin{aligned}
&\int_0^t g(t,s)|t-s|^{-\alpha}u(s)ds \\
&\quad = \int_0^t [g(t,s) - g(t,t)]|t-s|^{-\alpha}u(s)ds + \int_0^t g(t,t)|t-s|^{-\alpha}u(s)ds \\
&\quad = \int_0^t [g(t,s) - g(t,t)]|t-s|^{-\alpha}u(s)ds \\
&\qquad - g(t,t)\left[\int_0^t (u(t) - u(s))|t-s|^{-\alpha}ds - \int_0^t u(t)|t-s|^{-\alpha}ds\right].
\end{aligned}$$

Now,

$$\begin{aligned}
|\int_0^t [g(t,s) - g(t,t)]|t-s|^{-\alpha}u(s)ds| &\leq \int_0^t |g(t,s) - g(t,t)||t-s|^{-\alpha}|u(s)|ds \\
&\leq \int_0^t L_s|t-s|^{1-\alpha}|u(s)|ds \longrightarrow 0 \\
&\qquad \textit{as} \quad s \to t, \quad \textit{since} \quad 0 < \alpha < 1.
\end{aligned}$$

where L_s is Lipschitz constant.
Also we have

$$\begin{aligned}
|\int_0^t [u(t) - u(s)]|t-s|^{-\alpha}ds| &\leq \int_0^t |u(t) - u(s)||t-s|^{-\alpha}ds \\
&\leq |u'(\xi)| \int_0^t |t-s|^{1-\alpha}ds \\
&\qquad \textit{(applying the mean value theorem)} \\
&= |u'(\xi)|\frac{|t-s|^{2-\alpha}}{2-\alpha} \longrightarrow 0, \\
&\qquad \textit{since} \quad 0 < \alpha < 1 \quad \textit{and} \quad 0 < \xi < t.
\end{aligned}$$

We introduce the function $K_1(t,s)$ as

$$K_1(t,s) = \begin{cases} g(t,s)|t-s|^{-\alpha}, & t \neq s, \\ 0, & t = s. \end{cases}$$

Applying J^α on both sides of eq. (10.37), we have

$$u(t) = h(t) + J^\alpha \left[\lambda \mathcal{K}_1 u(t + \mu \mathcal{K}_2 u(t) + \mathcal{F}(t)\right], \tag{10.38}$$

where $h(t) = \sum_{k=0}^{n-1} \frac{t^k}{k!} u^{(k)}(0+), \quad n-1 < \alpha < n$. Eq. (10.38) can be written as fixed point equation form $\mathcal{A}u = u$, where $\mathcal{A}$ is defined as

$$\mathcal{A}u(t) = h(t) + J^\alpha \left[\lambda \mathcal{K}_1 u(t + \mu \mathcal{K}_2 u(t) + \mathcal{F}(t)\right]. \tag{10.39}$$

Let $(C[0,1], \|.\|_\infty)$ be the Hilbert space of all continuous functions with norm $\|f\|_\infty = \max_t |f(t)|$. Then we proceed to prove the uniqueness of the solution of the eq. (10.36).

Theorem 10.5.1. *If* $|\lambda|\|\mathcal{K}_1\|_\infty + |\mu|\|\mathcal{K}_2\|_\infty < \Gamma(\alpha+1)$, *then the initial value problem* (10.36) *has an unique solution* $u \in C[0,1]$.

Proof. Let $\mathcal{A} : C[0,1] \to C[0,1]$ such that

$$\mathcal{A}u(t) = h(t) + \frac{1}{\Gamma(\alpha)} \int_0^t (t-\tau)^{\alpha-1} \left[\lambda \mathcal{K}_1 u(\tau) + \mu \mathcal{K}_2 u(\tau) + \mathcal{F}(\tau)\right] d\tau.$$

Let $\tilde{u}, u \in C[0,1]$ and

$$\begin{aligned} \mathcal{A}\tilde{u}(t) - \mathcal{A}u(t) = \frac{1}{\Gamma(\alpha)} \int_0^t & (t-\tau)^{\alpha-1} \\ & \times \left[\lambda \mathcal{K}_1(\tilde{u}(\tau) - u(\tau)) + \mu \mathcal{K}_2(\tilde{u}(\tau) - u(\tau))\right] d\tau. \end{aligned}$$

Then for $t > 0$, we have

$$\begin{aligned} |\mathcal{A}\tilde{u}(t) - \mathcal{A}u(t)| &\leq \frac{1}{\Gamma(\alpha)} \int_0^t |t-\tau|^{\alpha-1} \\ &\quad \times \left[|\lambda||\mathcal{K}_1||\tilde{u}(\tau) - u(\tau)| + |\mu||\mathcal{K}_2||\tilde{u}(\tau) - u(\tau)|\right] d\tau \\ &\leq \frac{1}{\Gamma(\alpha)} \int_0^t |t-\tau|^{\alpha-1} (|\lambda|\|\mathcal{K}_1\|_\infty + |\mu|\|\mathcal{K}_2\|_\infty)\|\tilde{u} - u\|_\infty \\ &= (|\lambda|\|\mathcal{K}_1\|_\infty + |\mu|\|\mathcal{K}_2\|_\infty)\|\tilde{u} - u\|_\infty \frac{|t|^\alpha}{\Gamma(\alpha+1)} \\ &\leq (|\lambda|\|\mathcal{K}_1\|_\infty + |\mu|\|\mathcal{K}_2\|_\infty)\|\tilde{u} - u\|_\infty \frac{1}{\Gamma(\alpha+1)}. \end{aligned}$$

Therefore,

$$\|\mathcal{A}\tilde{u}(t) - \mathcal{A}u(t)\|_\infty \leq \Omega_{\mathcal{K}_1,\mathcal{K}_2,\lambda,\mu,\alpha}\|\tilde{u} - u\|_\infty,$$

where

$$\Omega_{\mathcal{K}_1,\mathcal{K}_2,\lambda,\mu,\alpha} = (|\lambda|\|\mathcal{K}_1\|_\infty + |\mu|\|\mathcal{K}_2\|_\infty)\frac{1}{\Gamma(\alpha+1)}.$$

Since $\Omega_{\mathcal{K}_1,\mathcal{K}_2,\lambda,\mu,\alpha} < 1$, by contraction mapping theorem, the initial value problem (10.36) has an unique solution in $C[0,1]$. □

10.5.2 Sinc basis function and its properties

In this section, we have discussed the notation and definition of Sinc functions, Sinc quadrature and Sinc function approximation.
The Sinc function is defined on the whole real line by [59]

$$Sinc(x) = \begin{cases} \frac{\sin(\pi x)}{\pi x}, & x \neq 0, \\ 1, & x = 0. \end{cases} \tag{10.40}$$

The sinc functions are translated by evenly spaced nodes as

$$S(k,h)(x) = Sinc\left(\frac{x-kh}{h}\right), \quad k = 0, \pm 1, \pm 2, ... \tag{10.41}$$

Let f be a function defined on the real line and $h > 0$; the series

$$C(f,h) = \sum_{k=-\infty}^{\infty} f(hk) Sinc\left(\frac{x-kh}{h}\right) \tag{10.42}$$

is called the Whittaker cardinal expansion of f whenever this series converges. Our aim is to construct the approximation on the interval (a,b), considering the conformal maps

$$\phi(z) = \ln\left(\frac{z-a}{b-z}\right). \tag{10.43}$$

The ϕ maps from an eye-shaped region

$$D_E = \{z = x + iy : |arg\left(\frac{z-a}{b-z}\right)| < d \leq \frac{\pi}{2}\}$$

onto the infinite strip

$$D_s = \{\zeta = \xi + i\eta : |\eta| < d \leq \frac{\pi}{2}\}.$$

This is shown in Figure 10.1.

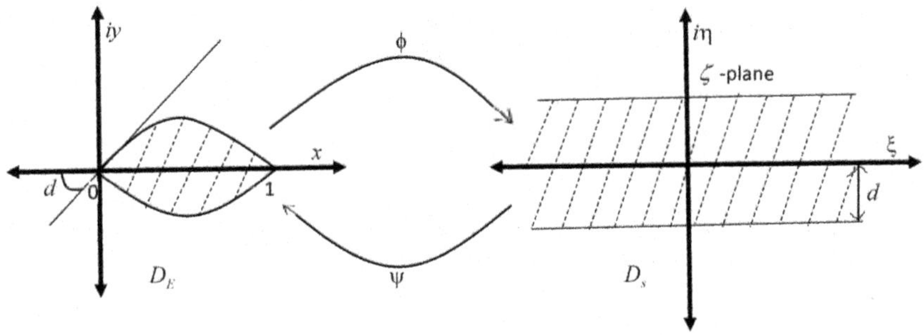

FIGURE 10.1: The domain D_E and D_s

The Sinc basis functions are derived from the composition

$$S_j(x) = S(j,h) \circ \phi(x) = sinc\left(\frac{\phi(x) - jh}{h}\right) \tag{10.44}$$

on the interval (a, b), where h is the mesh size in D_s. The Sinc grid points $x_k \in (a, b)$ in D_E will be defined as the inverse images of the equispaced grids as

$$x_k = \phi^{-1}(kh) = \frac{a + be^{kh}}{1 + e^{kh}}. \tag{10.45}$$

We define the range of ϕ^{-1} on the real line as

$$\Gamma = \{\psi(u) = \phi^{-1}(u) \in D_E : -\infty < u < \infty\} = (a, b).$$

Definition 10.5.1. [59] *Let $B(D_E)$ be the class of functions F that are analytic in D_E and satisfy*

$$\int_{\psi(L+u)} |F(z)dz| \to 0, \qquad as \;\; u = \pm\infty$$

where $L = \{iy : |y| < d \leq \frac{\pi}{2}$, and on the boundary of D_E (denoted ∂D_E) satisfy $T(F) = \int_{\partial D_E} |F(z)dz| < \infty$.

Theorem 10.5.2. [59] *Let Γ be (a, b), if $F \in B(D_E)$, then for $h > 0$ sufficiently small*

$$\int_\Gamma F(z)dz - h\sum_{j=-\infty}^{\infty} \frac{F(z_j)}{\phi'(z_j)} = \frac{i}{2}\int_{\partial D} \frac{F(z)k(\phi,h)(z)}{\sin(\pi\phi(z)/h)}dz \equiv I_F, \tag{10.46}$$

where

$$|k(\phi, h)|_{z\in\partial D} = |\exp\left[\frac{i\pi\phi(z)}{h}sgn(Im\phi(z))\right]|_{z\in\partial D} = e^{-\pi d/h}.$$

Theorem 10.5.3. [59, 190] *If there exist positive constants β, γ and C such that*

$$|\frac{F(x)}{\phi'(x)}| \leq C \begin{cases} \exp(-\beta|\phi(x)|), & x \in \psi((-\infty, 0)), \\ \exp(-\gamma|\phi(x)|), & x \in \psi((0, \infty)), \end{cases} \tag{10.47}$$

then the error bound for the quadrature rule is

$$|\int_\Gamma F(x)dx - h\sum_{j=-M}^{N} \frac{F(z_j)}{\phi'(z_j)}| \leq C\left(\frac{e^{-\beta M h}}{\beta} + \frac{e^{-\gamma N h}}{\gamma}\right) + |I_F|. \tag{10.48}$$

The infinite sum in eq. (10.46) is truncated with the use of eq. (10.47) to form the inequality (10.48) by choosing

$$h = \sqrt{\frac{\pi d}{\beta M}},$$

and

$$N \equiv \left[|\frac{\beta}{\gamma}M + 1|\right],$$

where $[x]$ is the integer part of x, then

$$\int_\Gamma F(x)dx = h\sum_{j=-M}^{N} \frac{F(z_j)}{\phi'(z_j)} + O(e^{-(\pi\beta d M)^{1/2}}). \tag{10.49}$$

Any function $f \in (0, 1)$ can be approximated by Sinc basis functions as

$$f_m(x) = \sum_{j=-N}^{N} c_j S_j(x), \quad m = 2N + 1, \tag{10.50}$$

where $S_j(x)$ is the composite function $S(j, h) \circ \phi(x)$ defined in eq. (10.44) and c_k's are the unknowns which can be determined.

10.5.3 Sinc-Galerkin method

Consider the problem defined in eq. (10.36) and approximate the unknown function $u(t)$ by Sinc basis functions defined in eq. (10.50). Taking the inner product with $S_k(t)$, we have the residual R_k as

$$\begin{aligned} R_k = \langle D_t^\alpha u(t), S_k(t)\rangle_{w(t)} - \lambda\langle \mathcal{K}_1 u(t), S_k(t)\rangle_{w(t)} - \mu\langle \mathcal{K}_2 u(t), S_k(t)\rangle_{w(t)} \\ - \langle f(t), S_k(t)\rangle_{w(t)}, \end{aligned} \tag{10.51}$$

where the inner-product is defined as

$$\langle f(t), g(t)\rangle_{w(t)} = \int_0^1 f(t)g(t)w(t)dt$$

and $w(t)$ is the weight function chosen as $w(t) = 1/\phi'(t)$.
Here, $\mathcal{K}_1 u(t)$ and $\mathcal{K}_2 u(t)$ are defined as follows

$$\mathcal{K}_1 u(t) = \int_0^t K_1(t,s)u(s)ds,$$
$$\mathcal{K}_2 u(t) = \int_0^1 K_2(t,s)u(s)ds.$$

We define the conformal map $\sigma(s)$ on the interval $(0,t)$ as

$$\sigma(s) = \ln\left(\frac{s}{t-s}\right)$$

and the inverse image is given by

$$s_r = \sigma^{-1}(rh_L) = \frac{te^s}{1+e^s}|_{s=rh_L},$$

where $h_L = \frac{\pi}{\sqrt{L}}$, $L \in \mathbb{Z}^+$.
Let us apply the Sinc quadrature defined in eq. (10.49) to the the followings:

$$\begin{aligned} D_t^\alpha u(t) &= \frac{1}{\Gamma(1-\alpha)}\int_0^t \frac{u'(s)}{(t-s)^\alpha}ds \\ &= \frac{h_L}{\Gamma(1-\alpha)}\sum_{r=-L}^{L}\frac{u'(s_r)}{(t-s_r)^\alpha \sigma'(s_r)} \end{aligned}$$

$$\begin{aligned} \langle D_t^\alpha u(t), S_k(t)\rangle_{w(t)} &= \int_0^1 D^\alpha u(t) S_k(t) w(t) dt \\ &= h\sum_{j=-N}^{N}\frac{D^\alpha u(t_j)S_k(t_j)w(t_j)}{\phi'(t_j)} \\ &= \frac{hh_L}{\Gamma(1-\alpha)}\sum_{r=-L}^{L}\frac{u'(s_r)w(t_k)}{(t_k-s_r)^\alpha \sigma'(s_r)\phi'(t_k)} \end{aligned} \tag{10.52}$$

Similarly,

$$\begin{aligned} \mathcal{K}_1 u(t) &= \int_0^t K_1(t,s)u(s)ds \\ &= h_L\sum_{r=-L}^{L}\frac{K_1(t,s_r)u(s_r)}{\sigma'(s_r)} \end{aligned}$$

$$\langle \mathcal{K}_1 u(t), S_k(t)\rangle_{w(t)} = \int_0^1 \mathcal{K}_1 u(t) S_k(t) w(t) dt$$
$$= h \sum_{j=-N}^{N} \frac{\mathcal{K}_1 u(t_j) S_k(t_j) w(t_j)}{\phi'(t_j)}$$
$$= h h_L \sum_{r=-L}^{L} \frac{K_1(t_k, s_r) u(s_r) w(t_k)}{\sigma'(s_r)\phi'(t_k)} \tag{10.53}$$

Again

$$\mathcal{K}_2 u(t) = \int_0^1 K_2(t,s) u(s) ds$$
$$= h \sum_{j=-N}^{N} \frac{K_2(t,t_j) u(t_j)}{\phi'(t_j)}$$
$$= h \sum_{j=-N}^{N} \frac{K_2(t,t_j) c_j}{\phi'(t_j)}$$

$$\langle \mathcal{K}_2 u(t), S_k(t)\rangle_{w(t)} = \int_0^1 \mathcal{K}_2 u(t) S_k(t) w(t) dt$$
$$= h \sum_{i=-N}^{N} \frac{\mathcal{K}_2 u(t_i) S_k(t_i) w(t_i)}{\phi'(t_i)}$$
$$= h^2 \sum_{j=-N}^{N} \frac{K_2(t_k, t_j) c_j w(t_k)}{\phi'(t_j)\phi'(t_k)} \tag{10.54}$$

and

$$\langle f(t), S_k(t)\rangle_{w(t)} = \int_0^1 f(t) S_k(t) w(t) dt$$
$$= h \sum_{j=-N}^{N} \frac{f(t_j) S_k(t_j) w(t_j)}{\phi'(t_j)}$$
$$= h \frac{f(t_k) w(t_k)}{\phi'(t_k)} \tag{10.55}$$

Using eqs. (10.52)-(10.55) in eq. (10.51), we have

$$R_k = \frac{h w(t_k)}{\phi'(t_k)} \left[\frac{h_L}{\Gamma(1-\alpha)} \sum_{r=-L}^{L} \frac{u'(s_r)}{(t_k - s_r)^{\alpha} \sigma'(s_r)} - \lambda h_L \sum_{r=-L}^{L} \frac{K_1(t_k, s_r) u(s_r)}{\sigma'(s_r)} \right.$$
$$\left. - \mu h \sum_{j=-N}^{N} \frac{K_2(t_k, t_j) c_j}{\phi'(t_j)} - f(t_k) \right]. \tag{10.56}$$

For $k = -N, ..., 0, ..., N$, the system (10.56) gives $2N + 1$ algebraic equations with the same number of unknowns as c_k, $k = -N, ..., 0, ..., N$. Solving this system numerically, we obtain the approximate value of c_k, $k = -N, ..., 0, ..., N$ and hence get the solution.

10.5.4 Error analysis

Theorem 10.5.4. *Assume that α, μ, d and β are uniquely defined and u, $K_1(t,s)$, $K_2(t,s)$ and f are analytic in D_E. Let ϕ be the conformal map from D_E onto D_s. The error in the solution of eq. (10.36) by applying the Sinc-Galerkin method is calculated as*

$$\|e\|_\infty \leq \left(4 + \frac{3}{|\phi'(kh)|^2}\right)\mathcal{M}e^{-\sqrt{\pi d\beta N}} + 3\mathcal{M}^2 e^{-2\sqrt{\pi d\beta N}}.$$

Proof. Consider the residual term defined in eq. (10.51) as

$$R_k = \langle D_t^\alpha u(t), S_k(t)\rangle_{w(t)} - \lambda\langle \mathcal{K}_1 u, S_k(t)\rangle_{w(t)} - \mu\langle \mathcal{K}_2 u, S_k(t)\rangle_{w(t)} - \langle f(t), S_k(t)\rangle_{w(t)}$$

We apply the Sinc quadrature rule for evaluation of fractional derivative D_t^α as

$$D_t^\alpha u(t) = \frac{h_L}{\Gamma(1-\alpha)}\sum_{r=-L}^{L}\frac{u'(s_r)}{(t-s_r)^\alpha \sigma'(s_r)} + L_1 e^{-\sqrt{\pi d\beta L}},$$

where L_1 is depending on the integrand and d, σ, D_E only.

$$\langle D_t^\alpha u(t), S_k(t)\rangle_{w(t)} = \frac{hh_L}{\Gamma(1-\alpha)}\sum_{r=-L}^{L}\frac{u'(s_r)w(t_k)}{(t_k-s_r)^\alpha \sigma'(s_r)\phi'(t_k)} + M_1 e^{-\sqrt{\pi d\beta N}} + \frac{1}{\phi'(kh)^2}L_1 e^{-\sqrt{\pi d\beta L}} + M_1 L_1 e^{-\sqrt{\pi d\beta L}}e^{-\sqrt{\pi d\beta N}},$$

where M_1 is depending on the integrand and d, ϕ, D_E only.
Similarly, we have

$$\langle \mathcal{K}_1 u(t), S_k(t)\rangle_{w(t)} = hh_L\sum_{r=-L}^{L}\frac{K_1(t_k,s_r)u(s_r)w(t_k)}{\sigma'(s_r)\phi'(t_k)} + M_2 e^{-\sqrt{\pi d\beta N}} + \frac{1}{\phi'(kh)^2}L_2 e^{-\sqrt{\pi d\beta L}} + M_2 L_2 e^{-\sqrt{\pi d\beta L}}e^{-\sqrt{\pi d\beta N}}$$

$$\langle \mathcal{K}_2 u(t), S_k(t)\rangle_{w(t)} = h^2\sum_{j=-N}^{N}\frac{K_2(t_k,t_j)c_j w(t_k)}{\phi'(t_j)\phi'(t_k)} + M_3 e^{-\sqrt{\pi d\beta N}} + \frac{1}{\phi'(kh)^2}M_3 e^{-\sqrt{\pi d\beta N}} + M_3^2 e^{-2\sqrt{\pi d\beta N}}$$

and

$$\langle f(t), S_k(t)\rangle_{w(t)} = h\frac{f(t_k)w(t_k)}{\phi'(t_k)} + M_4 e^{-\sqrt{\pi d\beta N}}.$$

The error term associated with those above calculations are collected as below by assuming $\mathcal{M} = \max\{M_1, M_2, M_3, M_4, L_1, L_2\}$, and $L = N$:

$$\|e\|_\infty \leq \left(4 + \frac{3}{|\phi'(kh)|^2}\right)\mathcal{M}e^{-\sqrt{\pi d\beta N}} + 3\mathcal{M}^2 e^{-2\sqrt{\pi d\beta N}}.$$

□

10.5.5 Illustrative examples

Example 10.5.1. *Let us consider the fractional Volterra-Fredholm integro-differential equation with weakly singular kernel [215]*

$$D_t^{0.25}u(t) = \frac{1}{2}\int_0^t \frac{u(s)}{(t-s)^{\frac{1}{2}}}ds + \frac{1}{3}\int_0^1 (t-s)u(s)ds + f(t),$$

with initial condition $u(0) = 0$. The exact solution of this problem is $t^2 + t^3$ and

$$f(t) = \frac{\Gamma(3)}{\Gamma(2.75)}t^{1.75} + \frac{\Gamma(4)}{\Gamma(3.75)}t^{2.75} - \frac{\sqrt{\pi}t^{5/2}\Gamma(3)}{2\Gamma(7/2)} - \frac{\sqrt{\pi}t^{7/2}\Gamma(4)}{2\Gamma(9/2)} - \frac{7}{36}t + \frac{3}{20}.$$

This problem has been solved by the Sinc-Galerkin method (for $N = 30$) which reduces the integral equation to a system of algebraic equations. Again the present solution compared with the solution obtained by CAS wavelet method [215]. The results obtained by the present method and their absolute errors have been shown in Table 10.7. Table 10.8 cites the comparison between the present method and the CAS wavelet method (CASWM).

Example 10.5.2. *Let us consider the fractional Volterra-Fredholm integro-differential equation with weak singular kernel [215]*

$$D_t^{0.15}u(t) = \frac{1}{4}\int_0^t \frac{u(s)}{(t-s)^{\frac{1}{2}}}ds + \frac{1}{7}\int_0^1 \exp(t+s)u(s)ds + f(t),$$

with initial condition $u(0) = 0$. The exact solution of this problem is $t(t-1)$ and

$$f(t) = \frac{\Gamma(3)}{\Gamma(2.85)}t^{1.85} - \frac{\Gamma(2)}{\Gamma(1.85)}t^{0.85} - \frac{\sqrt{\pi}t^{5/2}\Gamma(3)}{4\Gamma(7/2)} + \frac{\sqrt{\pi}t^{3/2}\Gamma(2)}{4\Gamma(5/2)} - \frac{e^{t+1} - 3e^t}{7}.$$

TABLE 10.7: Numerical results for Example 10.5.1

x	Exact	SGM	Error
0.1	0.011	0.0131253	2.12527E-3
0.2	0.048	0.0470762	9.23786E-4
0.3	0.117	0.117428	4.2783E-4
0.4	0.224	0.227391	3.39123E-3
0.5	0.375	0.374651	3.48864E-4
0.6	0.576	0.570188	5.8124E-3
0.7	0.833	0.831297	1.7027E-3
0.8	1.152	1.14896	3.04066E-3
0.9	1.539	1.54229	3.29448E-3

TABLE 10.8: Comparison of numerical results between SGM and CASWM for Example 10.5.1

x	Absolute error by	
	SGM	CASWM
1/6	1.47244E-3	1.1460E-2
2/6	1.41785E-3	9.6982E-3
3/6	3.48864E-4	2.9504E-3
4/6	3.49937E-3	9.6733E-3
5/6	2.4408E-3	2.9484E-2

This problem has been solved by the Sinc-Galerkin method (for $N = 20$) which reduces the integral equation to a system of algebraic equations. Again the present solution is compared with the solution obtained by the CAS wavelet method [215]. The results obtained by the present method and their absolute errors have been shown in Table 10.9. Table 10.10 cites the comparison between the present method and the CAS wavelet method (CASWM).

Example 10.5.3. *As the third example, let us consider the fractional Volterra integro-differential equation with weakly singular kernel [228]*

$$D_t^{\frac{1}{3}} u(t) = f(t) + p(t)u(t) + \int_0^t (t-s)^{-\frac{1}{2}} u(s)ds,$$

with initial condition $u(0) = 0$ and

$$p(t) = -\frac{32}{35} t^{1/2},$$

$$f(t) = \frac{8\pi t}{9\sqrt{3}\Gamma(2/3)} + \frac{81 t^{8/3}}{40\Gamma(2/3)} + t^{11/6}\left(\frac{32}{35} - \frac{\sqrt{\pi}\Gamma(7/3)}{\Gamma(17/6)}\right).$$

The exact solution of this problem is $t^3 + t^{4/3}$. This problem has been solved by the Sinc-Galerkin method (for $N = 30$) which reduces the integral equation

TABLE 10.9: Numerical results for Example 10.5.2

x	Exact	SGM	Error
0.1	-0.09	-0.0899617	3.83209E-5
0.2	-0.16	-0.159942	5.84668E-5
0.3	-0.21	-0.209918	8.19411E-5
0.4	-0.24	-0.23989	1.10007E-4
0.5	-0.25	-0.249866	1.34402E-4
0.6	-0.24	-0.239847	1.52509E-4
0.7	-0.21	-0.209828	1.7195E-4
0.8	-0.16	-0.159818	1.81826E-4
0.9	-0.09	-0.0898167	1.83312E-4

TABLE 10.10: Comparison of numerical results between SGM and CASWM for Example 10.5.2

x	Exact	Numerical results by		Absolute error by	
		SGM	CASWM	SGM	CASWM
1/6	-0.138889	-0.138837	-0.133387	5.14283E-5	5.50189E-3
2/6	-0.222222	-0.222131	-0.215341	9.10148E-5	6.79021E-3
3/6	-0.25	-0.249866	-0.242660	1.34402E-4	7.2056E-3
4/6	-0.222222	-0.222057	-0.215141	1.65325E-4	6.9159E-3
5/6	-0.138889	-0.138709	-0.132694	1.80252E-4	6.01464E-3

to a system of algebraic equations. The results obtained by the present method and their absolute errors have been shown in Table 10.11.

10.6 Conclusion

In this chapter, we have employed efficient numerical techniques for solving fractional integro-differential equations. Especially, Galerkin techniques are used to solve these equations. These methods reduce the integral equations to a system of nonlinear algebraic equations and that algebraic system has been solved by Newton's method. The Gauss-Legendre quadrature has been applied for evaluating the integrations that arise in the problems. In Section 10.3, the Legendre wavelet method has been applied to solve nonlinear fractional Volterra-Fredholm integro-differential equations with mixed boundary conditions. Also the obtained results again compare with the results obtained by the Nystrom method which is available in the literature. Furthermore, we can get significantly less absolute error by increasing the order of Legendre polynomials. In Section 10.4, the Legendre wavelet Petrov-Galerkin method has been applied to solve fractional Volterra integro-differential equa-

TABLE 10.11: Numerical results for Example 10.5.3

x	Exact	SGM	Error
0.1	0.0474159	0.0491855	1.76957E-3
0.2	0.124961	0.1248	1.60604E-4
0.3	0.22783	0.227312	5.1822E-4
0.4	0.358723	0.361514	2.79194E-3
0.5	0.52185	0.522219	3.69227E-4
0.6	0.72206	0.717003	5.05652E-3
0.7	0.964533	0.963979	5.53609E-4
0.8	1.25465	1.25161	3.04882E-3
0.9	1.59794	1.60088	2.94377E-3

tions and the obtained results are in good agreement with the exact results. In Section 10.5, the Sinc-Galerkin method has been applied to solve fractional Fredholm-Volterra integro-differential equations with weakly singular kernels. The present method results have been compared with the results obtained by the CAS wavelet method. The illustrative examples have been included to demonstrate the validity and applicability of the proposed techniques. Undoubtedly these examples also exhibit the accuracy and efficiency of the present methods.

Bibliography

[1] S. Abbasbandy, E. Babolian, and M. Alavi. Numerical method for solving linear Fredholm fuzzy integral equations of the second kind. *Chaos, Solitons & Fractals*, 31(1):138–146, 2007.

[2] S. Abbasbandy and M. S. Hashemi. A series solution of fuzzy integro-differential equations. *Journal of Fuzzy Set Valued Analysis*, 2012, Article ID: jfsva–00066, 2012.

[3] W. M. Abd-Elhameed, E. H. Doha, and Y. H. Youssri. New spectral second kind Chebyshev wavelets algorithm for solving linear and nonlinear second-order differential equations involving singular and Bratu type equations. *Abstract and Applied Analysis*, 2013, Article ID: 715756, 2013.

[4] W. M. Abd-Elhameed, E. H. Doha, and Y. H. Youssri. New wavelets collocation method for solving second-order multipoint boundary value problems using Chebyshev polynomials of third and fourth kinds. *Abstract and Applied Analysis*, 2013, Article ID: 542839, 2013.

[5] Hojatollah Adibi and Pouria Assari. Chebyshev wavelet method for numerical solution of Fredholm integral equations of the first kind. *Mathematical problems in Engineering*, 2010, Article ID: 138408, 2010.

[6] Yousef Al-Jarrah and En-Bing Lin. Wavelet interpolation method for solving singular integral equations. *Applied Mathematics*, 4(11):1–4, 2013.

[7] Ayyaz Ali, Muhammad Asad Iqbal, and Syed Tauseef Mohyud-Din. Chebyshev wavelets method for boundary value problems. *Scientific Research and Essays*, 8(46):2235–2241, 2013.

[8] Robab Alikhani, Fariba Bahrami, and Adel Jabbari. Existence of global solutions to nonlinear fuzzy Volterra integro-differential equations. *Nonlinear Analysis: Theory, Methods & Applications*, 75(4):1810–1821, 2012.

[9] Sertan Alkan and Aydin Secer. Application of Sinc-Galerkin method for solving space-fractional boundary value problems. *Mathematical Problems in Engineering*, 2015, Article ID: 217348, 2015.

[10] T. Allahviranloo and S. Hashemzehi. The homotopy perturbation method for fuzzy Fredholm integral equations. *Journal of Applied Mathematics, Islamic Azad University of Lahijan*, 5:1–12, 2008.

[11] Hossein Aminikhah and Sakineh Moradian. Numerical solution of singular Lane-Emden equation. *ISRN Mathematical Physics*, 2013, Article ID: 507145, 2013.

[12] M. A. Fariborzi Araghi and Nouredin Parandin. Numerical solution of fuzzy Fredholm integral equations by the Lagrange interpolation based on the extension principle. *Soft Computing*, 15(12):2449–2456, 2011.

[13] Aytac Arikoglu and Ibrahim Ozkol. Solution of differential–difference equations by using differential transform method. *Applied Mathematics and Computation*, 181(1):153–162, 2006.

[14] Aytac Arikoglu and Ibrahim Ozkol. Solution of fractional integro-differential equations by using fractional differential transform method. *Chaos, Solitons & Fractals*, 40(2):521–529, 2009.

[15] B. Asady, M. Tavassoli Kajani, A. Hadi Vencheh, and A. Heydari. Direct method for solving integro differential equations using hybrid Fourier and block-pulse functions. *International Journal of Computer Mathematics*, 82(7):889–895, 2005.

[16] M. Asgari. Block pulse approximation of fractional stochastic integro-differential equation. *Communications in Numerical Analysis*, 2014, Article ID: cna-00212:1–7, 2014.

[17] B. Babayar-Razlighi and Babak Soltanalizadeh. Numerical solution for system of singular nonlinear Volterra integro-differential equations by Newton-Product method. *Applied Mathematics and Computation*, 219(15):8375–8383, 2013.

[18] B. Babayar-Razlighi and Babak Soltanalizadeh. Numerical solution of a nonlinear singular Volterra integral system by the Newton product integration method. *Mathematical and Computer Modelling*, 58(11):1696–1703, 2013.

[19] E. Babolian and J. Biazar. Solution of a system of nonlinear Volterra integral equations of the second kind. *Far East Journal of Mathematical Sciences*, 2(6):935–946, 2000.

[20] E. Babolian, F. Fattahzadeh, and E. Golpar Raboky. A Chebyshev approximation for solving nonlinear integral equations of Hammerstein type. *Applied Mathematics and Computation*, 189(1):641–646, 2007.

[21] Esmail Babolian and J. Biazar. Solving the problem of biological species living together by Adomian decomposition method. *Applied Mathematics and Computation*, 129(2):339–343, 2002.

[22] Esmail Babolian, J. Biazar, and A. R. Vahidi. The decomposition method applied to systems of Fredholm integral equations of the second kind. *Applied Mathematics and Computation*, 148(2):443–452, 2004.

[23] Esmail Babolian, H. Sadeghi Goghary, and Saeid Abbasbandy. Numerical solution of linear Fredholm fuzzy integral equations of the second kind by Adomian method. *Applied Mathematics and Computation*, 161(3):733–744, 2005.

[24] Esmail Babolian, Zahra Masouri, and Saeed Hatamzadeh-Varmazyar. A direct method for numerically solving integral equations system using orthogonal triangular functions. *International Journal of Industrial Mathematics*, 1(2):135–145, 2009.

[25] Mahdy Baghmisheh and Reza Ezzati. Numerical solution of nonlinear fuzzy Fredholm integral equations of the second kind using hybrid of block-pulse functions and Taylor series. *Advances in Difference Equations*, 2015(1):1–15, 2015.

[26] Ronald L. Bagley and P. J. Torvik. A theoretical basis for the application of fractional calculus to viscoelasticity. *Journal of Rheology*, 27(3):201–210, 1983.

[27] Richard T. Baillie. Long memory processes and fractional integration in econometrics. *Journal of Econometrics*, 73(1):5–59, 1996.

[28] Tamal Banerjee and Ashok Khanna. Infinite dilution activity coefficients for trihexyltetradecyl phosphonium ionic liquids: measurements and COSMO-RS prediction. *Journal of Chemical & Engineering Data*, 51(6):2170–2177, 2006.

[29] E. Banifatemi, M. Razzaghi, and S. Yousefi. Two-dimensional Legendre wavelets method for the mixed Volterra-Fredholm integral equations. *Journal of Vibration and Control*, 13(11):1667–1675, 2007.

[30] M. Barkhordari Ahmadi and M. Khezerloo. Fuzzy bivariate Chebyshev method for solving fuzzy Volterra-Fredholm integral equations. *International Journal of Industrial Mathematics*, 3(2):67–77, 2011.

[31] Behrooz Basirat and Mohammad Amin Shahdadi. Numerical solution of nonlinear integro-differential equations with initial conditions by Bernstein operational matrix of derivative. *International Journal of Modern Nonlinear Theory and Application*, 2(02):141–149, 2013.

[32] Gregory Beylkin, Ronald Coifman, and Vladimir Rokhlin. Fast wavelet transforms and numerical algorithms I. *Communications on Pure and Applied Mathematics*, 44(2):141–183, 1991.

[33] Dambaru D. Bhatta and Muhammad I. Bhatti. Numerical solution of KdV equation using modified Bernstein polynomials. *Applied Mathematics and Computation*, 174(2):1255–1268, 2006.

[34] Subhra Bhattacharya and B. N. Mandal. Numerical solution of a singular integro-differential equation. *Applied Mathematics and Computation*, 195(1):346–350, 2008.

[35] Subhra Bhattacharya and B. N. Mandal. Use of Bernstein polynomials in numerical solutions of Volterra integral equations. *Applied Mathematical Sciences*, 2(33-36):1773–1787, 2008.

[36] M. Idrees Bhatti and P. Bracken. Solutions of differential equations in a Bernstein polynomial basis. *Journal of Computational and Applied Mathematics*, 205(1):272–280, 2007.

[37] J. Biazar, Z. Ayati, and M. Partovi. Homotopy perturbation method for biological species living together. *International Journal of Applied*, 2(1):44–48, 2013.

[38] J. Biazar and H. Ebrahimi. Legendre wavelets for systems of Fredholm integral equations of the second kind. *World Applied Sciences Journal*, 9(9):1008–1012, 2010.

[39] J. Biazar, H. Ghazvini, and M. Eslami. He's homotopy perturbation method for systems of integro-differential equations. *Chaos, Solitons & Fractals*, 39(3):1253–1258, 2009.

[40] Jafar Biazar and Hamideh Ebrahimi. A strong method for solving systems of integro-differential equations. *Applied Mathematics*, 2(09):1105–1113, 2011.

[41] Alexandru Mihai Bica. One-sided fuzzy numbers and applications to integral equations from epidemiology. *Fuzzy Sets and Systems*, 219:27–48, 2013.

[42] Alexandru Mihai Bica and Constantin Popescu. Numerical solutions of the nonlinear fuzzy Hammerstein–Volterra delay integral equations. *Information Sciences*, 223:236–255, 2013.

[43] Alexandru Mihai Bica and Constantin Popescu. Approximating the solution of nonlinear Hammerstein fuzzy integral equations. *Fuzzy Sets and Systems*, 245:1–17, 2014.

[44] G. W. Scott Blair and J. E. Caffyn. An application of the theory of quasi-properties to the treatment of anomalous strain-stress relations. *Philosophical Magazine*, 40(300):80–94, 1949.

[45] Tian-Li Bo, Li Xie, and Xiao Jing Zheng. Numerical approach to wind ripple in desert. *International Journal of Nonlinear Sciences and Numerical Simulation*, 8(2):223–228, 2007.

[46] M. Caputo. *Elasticita e Dissipazione*. Zanichelli, Bologna, 1969.

[47] M. Caputo and F. Mainardi. A new dissipation model based on memory mechanism. *Pure and Applied Geophysics*, 91(1):134–147, 1971.

[48] Michele Caputo. Linear models of dissipation whose Q is almost frequency independent-II. *Geophysical Journal of the Royal Astronomical Society*, 13(5):529–539, 1967.

[49] Subrahmanyan Chandrasekhar. *An Introduction to the Study of Stellar Structure*. Dover, New York, 1967.

[50] Sheldon S. L. Chang and Lofti A. Zadeh. On fuzzy mapping and control. *IEEE Transactions on Systems, Man and Cybernetics*, (1):30–34, 1972.

[51] Yanping Chen and Tao Tang. Spectral methods for weakly singular Volterra integral equations with smooth solutions. *Journal of Computational and Applied Mathematics*, 233(4):938–950, 2009.

[52] Zhongying Chen and Yuesheng Xu. The Petrov–Galerkin and iterated Petrov–Galerkin methods for second-kind integral equations. *SIAM Journal on Numerical Analysis*, 35(1):406–434, 1998.

[53] Charles K. Chui. *An Introduction to Wavelets, Wavelet Analysis and Its Applications*, volume 1. Academic Press, Massachusetts, 1992.

[54] Charles K. Chui. *Wavelets: A Mathematical Tool for Signal Analysis*, volume 1. SIAM, 1997.

[55] Kenneth L. Cooke. An epidemic equation with immigration. *Mathematical Biosciences*, 29(1):135–158, 1976.

[56] Chaoqing Dai and Jiefang Zhang. Jacobian elliptic function method for nonlinear differential-difference equations. *Chaos, Solitons & Fractals*, 27(4):1042–1047, 2006.

[57] Harold Thayer Davis. *Introduction to Nonlinear Differential and Integral Equations*. Dover, New York, 1962.

[58] E. H. Doha, W. M. Abd-Elhameed, and Y. H. Youssri. Second kind Chebyshev operational matrix algorithm for solving differential equations of Lane–Emden type. *New Astronomy*, 23:113–117, 2013.

[59] M. El-Gamel and A. I. Zayed. Sinc-Galerkin method for solving nonlinear boundary-value problems. *Computers & Mathematics with Applications*, 48(9):1285–1298, 2004.

[60] R. Ezzati and F. Mokhtari. Numerical solution of Fredholm integral equations of the second kind by using fuzzy transforms. *International Journal of Physical Science*, 7:1578–1583, 2012.

[61] R. Ezzati and S. Ziari. Numerical solution and error estimation of fuzzy Fredholm integral equation using fuzzy Bernstein polynomials. *Australian Journal of Basic and Applied Sciences*, 5(9):2072–2082, 2011.

[62] Hadi Hosseini Fadravi, Reza Buzhabadi, and Hassan Saberi Nik. Solving linear Fredholm fuzzy integral equations of the second kind by artificial neural networks. *Alexandria Engineering Journal*, 53(1):249–257, 2014.

[63] Robert Franke and Bernd Hannebauer. On the influence of basis sets and quantum chemical methods on the prediction accuracy of COSMO-RS. *Physical Chemistry Chemical Physics*, 13(48):21344–21350, 2011.

[64] Shou-Zhong Fu, Zhong Wang, and Jun-Sheng Duan. Solution of quadratic integral equations by the Adomian decomposition method. *CMES-Comput. Model. Eng. Sci*, 92(4):369–385, 2013.

[65] Nizami Gasilov, Şahin Emrah Amrahov, and Afet Golayoglu Fatullayev. Solution of linear differential equations with fuzzy boundary values. *Fuzzy Sets and Systems*, 257:169–183, 2014.

[66] Hubert Goenner and Peter Havas. Exact solutions of the generalized Lane-Emden equation. *Journal of Mathematical Physics*, 41:7029–7042, 2000.

[67] Roy Goetschel and William Voxman. Elementary fuzzy calculus. *Fuzzy Sets and Systems*, 18(1):31–43, 1986.

[68] Jaideva C. Goswami and Andrew K. Chan. *Fundamentals of Wavelets: Theory, Algorithms, and Applications*, volume 233. John Wiley & Sons, New Jersey, 2011.

[69] Jaideva C. Goswami, Andrew K. Chan, and Charles K. Chui. On solving first-kind integral equations using wavelets on a bounded interval. *IEEE Transactions on Antennas and Propagation*, 43(6):614–622, 1995.

[70] Hans Grensemann and Jürgen Gmehling. Performance of a conductor-like screening model for real solvents model in comparison to classical group contribution methods. *Industrial & Engineering Chemistry Research*, 44(5):1610–1624, 2005.

[71] Mustafa Gülsu, Yalçın Öztürk, and Mehmet Sezer. A new collocation method for solution of mixed linear integro-differential-difference equations. *Applied Mathematics and Computation*, 216(7):2183–2198, 2010.

[72] Mustafa Gülsu and Mehmet Sezer. Approximations to the solution of linear Fredholm integro–differential–difference equation of high order. *Journal of the Franklin Institute*, 343(7):720–737, 2006.

[73] Mustafa Gülsu, Mehmet Sezer, and Bekir Tanay. A matrix method for solving high-order linear difference equations with mixed argument using hybrid Legendre and Taylor polynomials. *Journal of the Franklin Institute*, 343(6):647–659, 2006.

[74] A. K. Gupta and S. Saha Ray. Traveling wave solution of fractional KdV-Burger-Kuramoto equation describing nonlinear physical phenomena. *AIP Advances*, 4(9):097120–1–11, 2014.

[75] Burcu Gürbüz, Mustafa Gülsu, and Mehmet Sezer. Numerical approach of high-order linear delay difference equations with variable coefficients in terms of Laguerre polynomials. *Mathematical and Computational Applications*, 16(1):267, 2011.

[76] Burcu Gürbüz, Mehmet Sezer, and Coşkun Güler. Laguerre collocation method for solving Fredholm integro-differential equations with functional arguments. *Journal of Applied Mathematics*, 2014, Article ID: 682398, 2014.

[77] Huilei Han, Yaping Liu, T. Lau, and Xiaoming He. New algorithm for the system of nonlinear weakly singular Volterra integral equations of the second kind and integro-differential equations. *J. Inf. Comput. Sci*, 7:1229–1235, 2010.

[78] Robert F. Heinemann and Aubrey B. Poore. Multiplicity, stability, and oscillatory dynamics of the tubular reactor. *Chemical Engineering Science*, 36(8):1411–1419, 1981.

[79] Robert F. Heinemann and Aubrey B. Poore. The effect of activation energy on tubular reactor multiplicity. *Chemical Engineering Science*, 37(1):128–131, 1982.

[80] E. Hesameddini and E. Asadolahifard. Solving systems of linear Volterra integro-differential equations by using Sinc-collocation method. *International Journal of Mathematical Engineering and Science*, 2(7):1–9, 2013.

[81] Esmail Hesameddini and Elham Asadollahifard. Numerical solution of multi-order fractional differential equations via the sinc collocation method. *Iranian Journal of Numerical Analysis and Optimization*, 5(1):37–48, 2015.

[82] Edyta Hetmaniok, Damian Słota, Tomasz Trawiński, and Roman Wituła. Usage of the homotopy analysis method for solving the nonlinear and linear integral equations of the second kind. *Numerical Algorithms*, 67(1):163–185, 2014.

[83] M. H. Heydari, M. R. Hooshmandasl, C. Cattani, and Ming Li. Legendre wavelets method for solving fractional population growth model in a closed system. *Mathematical Problems in Engineering*, 2013, Article ID: 161030, 2013.

[84] M. H. Heydari, M. R. Hooshmandasl, F. Mohammadi, and C. Cattani. Wavelets method for solving systems of nonlinear singular fractional Volterra integro-differential equations. *Communications in Nonlinear Science and Numerical Simulation*, 19(1):37–48, 2014.

[85] M. S. Ismail, Farida M. Mosally, and Khadeejah M. Alamoudi. Petrov-Galerkin method for the coupled Schrödinger-KdV equation. *Abstract and Applied Analysis*, 2014, Article ID: 705204, 2014.

[86] Ahmad Jafarian, Safa A. Measoomy Nia, Alireza K. Golmankhaneh, and Dumitru Baleanu. Numerical solution of linear integral equations system using the Bernstein collocation method. *Advances in Difference Equations*, 2013(1):1–15, 2013.

[87] M. Jahantigh, T. Allahviranloo, and M. Otadi. Numerical solution of fuzzy integral equation. *Appl. Math. Sci.*, 2:33–46, 2008.

[88] Husein Jaradat, Fadi Awawdeh, and E. A. Rawashdeh. Analytic solution of fractional integro-differential equations. *Annals of the University of Craiova-Mathematics and Computer Science Series*, 38(1):1–10, 2011.

[89] Abdul Jerri. *Introduction to Integral Equations with Applications*. John Wiley & Sons, New York, 1999.

[90] Osmo Kaleva. Fuzzy differential equations. *Fuzzy Sets and Systems*, 24(3):301–317, 1987.

[91] Hideaki Kaneko, Richard D. Noren, and Boriboon Novaprateep. Wavelet applications to the Petrov–Galerkin method for Hammerstein equations. *Applied Numerical Mathematics*, 45(2):255–273, 2003.

[92] Seydi Battal Gazi Karakoç, Aytekin Eryılmaz, and Musa Başbük. The approximate solutions of Fredholm integro–differential-difference equations with variable coefficients via homotopy analysis method. *Mathematical Problems in Engineering*, 2013, Article ID: 261645, 2013.

[93] Daria Karpenko, Robert A. Van Gorder, and Abraham Kandel. The Cauchy problem for complex fuzzy differential equations. *Fuzzy Sets and Systems*, 245:18–29, 2014.

[94] Jun Kawabe. The bounded convergence in measure theorem for nonlinear integral functionals. *Fuzzy Sets and Systems*, 271:31–42, 2015.

[95] M. A. Kelmanson and M. C. Tenwick. Error reduction in Gauss-Jacobi-Nyström quadrature for Fredholm integral equations of the second kind. *Computer Modeling in Engineering and Sciences (CMES)*, 55(2):191, 2010.

[96] E. Keshavarz, Y. Ordokhani, and M. Razzaghi. Bernoulli wavelet operational matrix of fractional order integration and its applications in solving the fractional order differential equations. *Applied Mathematical Modelling*, 38(24):6038–6051, 2014.

[97] Mohammad Keyanpour and Tahereh Akbarian. New approach for solving of linear Fredholm fuzzy integral equations using sinc function. *J. of Mathematics and Computer Science*, 3:422–431, 2011.

[98] Masood C. Khalique, Fazal M. Mahomed, and Ben Muatjetjeja. Lagrangian formulation of a generalized Lane-Emden equation and double reduction. *Journal of Nonlinear Mathematical Physics*, 15(2):152–161, 2008.

[99] Yasir Khan, Zdeněk Svoboda, and Zdeněk Šmarda. Solving certain classes of Lane-Emden type equations using the differential transformation method. *Advances in Difference Equations*, 2012(1):1–11, 2012.

[100] Andreas Klamt. Conductor-like screening model for real solvents: a new approach to the quantitative calculation of solvation phenomena. *The Journal of Physical Chemistry*, 99(7):2224–2235, 1995.

[101] Andreas Klamt, Volker Jonas, Thorsten Bürger, and John C. W. Lohrenz. Refinement and parametrization of COSMO-RS. *The Journal of Physical Chemistry A*, 102(26):5074–5085, 1998.

[102] Omran Kouba. Lecture notes, Bernoulli polynomials and applications. *arXiv preprint arXiv:1309.7560*, 2013.

[103] Ayşe Kurt, Salih Yalçınbaş, and Mehmet Sezer. Fibonacci collocation method for solving high-order linear Fredholm integro-differential-difference equations. *International Journal of Mathematics and Mathematical Sciences*, 2013, Article ID: 486013, 2013.

[104] Muhammet Kurulay and Aydin Secer. Variational iteration method for solving nonlinear fractional integro-differential equations. *Int. J. of Computer Science and Emerging Technologies*, 2(1):18–20, 2011.

[105] Young-Chel Kwun, Mi-Ju Kim, Bu-Young Lee, and Jin-Han Park. Existence of solutions for the semilinear fuzzy integro-differential equations using by successive iteration. *Journal of Korean Institute of Intelligent Systems*, 18(4):543–548, 2008.

[106] Young Chel Kwun, Mi Ju Kim, Jong Seo Park, and Jin Han Park. Continuously initial observability for the semilinear fuzzy integrodifferential equations. In *Fifth International Conference on Fuzzy Systems and Knowledge Discovery*, volume 1, pages 225–229. IEEE, 2008.

[107] M. Lakestani, M. Razzaghi, and M. Dehghan. Solution of nonlinear Fredholm-Hammerstein integral equations by using semiorthogonal spline wavelets. *Mathematical Problems in Engineering*, 2005(1):113–121, 2005.

[108] M. Lakestani, M. Razzaghi, and M. Dehghan. Semiorthogonal spline wavelets approximation for Fredholm integro-differential equations. *Mathematical Problems in Engineering*, 2006, Article ID: 96184, 2006.

[109] Mehrdad Lakestani, Behzad Nemati Saray, and Mehdi Dehghan. Numerical solution for the weakly singular Fredholm integro-differential equations using Legendre multiwavelets. *Journal of Computational and Applied Mathematics*, 235(11):3291–3303, 2011.

[110] L. J. Lardy. A variation of Nystrom method for Hammerstein equations. *Journal of Integral Equations*, 3(1):43–60, 1981.

[111] Stig Larsson, Milena Racheva, and Fardin Saedpanah. Discontinuous Galerkin method for an integro-differential equation modeling dynamic fractional order viscoelasticity. *Computer Methods in Applied Mechanics and Engineering*, 283:196–209, 2015.

[112] Tao Lin, Yanping Lin, Ming Rao, and Shuhua Zhang. Petrov-Galerkin methods for linear Volterra integro-differential equations. *SIAM Journal on Numerical Analysis*, 38(3):937–963, 2001.

[113] Xiao-Yan Lin, Jin-Song Leng, and Ying-Jin Lu. A Haar wavelet solution to Fredholm equations. In *Proceedings of the International Conference on Computational Intelligence and Software Engineering*, pages 1–4, 2009.

[114] Chein-Shan Liu and Satya N. Atluri. A fictitious time integration method for the numerical solution of the Fredholm integral equation and for numerical differentiation of noisy data, and its relation to the filter theory. *Computer Modeling in Engineering and Sciences (CMES)*, 41(3):243, 2009.

[115] Nanshan Liu and En-Bing Lin. Legendre wavelet method for numerical solutions of partial differential equations. *Numerical methods for partial differential equations*, 26(1):81–94, 2010.

[116] Taher Lotfi and Katayoun Mahdiani. Fuzzy Galerkin method for solving Fredholm integral equations with error analysis. *International Journal of Industrial Mathematics*, 3(4):237–249, 2011.

[117] Heping Ma and Weiwei Sun. Optimal error estimates of the Legendre–Petrov–Galerkin method for the Korteweg–de Vries equation. *SIAM Journal on Numerical Analysis*, 39(4):1380–1394, 2001.

[118] Patryk Mach. All solutions of the n= 5 Lane–Emden equation. *Journal of Mathematical Physics*, 53(6):062503, 2012.

[119] N. M. Madbouly, D. F. McGhee, and G. F. Roach. Adomian's method for Hammerstein integral equations arising from chemical reactor theory. *Applied Mathematics and Computation*, 117(2):241–249, 2001.

[120] Y. Mahmoudi. Wavelet Galerkin method for numerical solution of nonlinear integral equation. *Applied Mathematics and Computation*, 167(2):1119–1129, 2005.

[121] Francesco Mainardi. Fractional calculus: some basic problems in continuum and statistical mechanics. *Fractals and Fractional Calculus in Continuum Mechanics*, 378:291–348, 1997.

[122] Hassan Majidian. Modified Euler's method with a graded mesh for a class of Volterra integral equations with weakly singular kernel. *Numerical Algorithms*, 67(2):405–422, 2014.

[123] K. Maleknejad, E. Hashemizadeh, and B. Basirat. Computational method based on Bernstein operational matrices for nonlinear Volterra–Fredholm–Hammerstein integral equations. *Communications in Nonlinear Science and Numerical Simulation*, 17(1):52–61, 2012.

[124] K. Maleknejad and M. Tavassoli Kajani. Solving second kind integral equations by Galerkin methods with hybrid Legendre and Block-Pulse functions. *Applied Mathematics and Computation*, 145(2):623–629, 2003.

[125] K. Maleknejad, M. Tavassoli Kajani, and Y. Mahmoudi. Numerical solution of linear Fredholm and Volterra integral equation of the second kind by using Legendre wavelets. *Kybernetes*, 32(9/10):1530–1539, 2003.

[126] Khosrow Maleknejad, N. Aghazadeh, and Mohsen Rabbani. Numerical solution of second kind Fredholm integral equations system by using a Taylor-series expansion method. *Applied Mathematics and Computation*, 175(2):1229–1234, 2006.

[127] Khosrow Maleknejad and Mahdiyeh Alizadeh. An efficient numerical scheme for solving Hammerstein integral equation arisen in chemical phenomenon. *Procedia Computer Science*, 3:361–364, 2011.

[128] Khosrow Maleknejad, Behrooz Basirat, and Elham Hashemizadeh. A Bernstein operational matrix approach for solving a system of high order linear Volterra–Fredholm integro-differential equations. *Mathematical and Computer Modelling*, 55(3):1363–1372, 2012.

[129] Khosrow Maleknejad, Elham Hashemizadeh, and Behrooz Basirat. Numerical solvability of Hammerstein integral equations based on hybrid Legendre and Block-Pulse functions. In *PDPTA*, pages 172–175, 2010.

[130] Khosrow Maleknejad and Farshid Mirzaee. Numerical solution of linear Fredholm integral equations system by rationalized Haar functions method. *International Journal of Computer Mathematics*, 80(11):1397–1405, 2003.

[131] Khosrow Maleknejad, M. Mohsenyzadeh, and Elham Hashemizadeh. Hybrid orthonormal Bernstein and Block-Pulse functions for solving Fredholm integral equations. In *Proceedings of the World Congress on Engineering*, volume 1, 2013.

[132] Khosrow Maleknejad and K. Nedaiasl. Application of sinc-collocation method for solving a class of nonlinear Fredholm integral equations. *Computers & Mathematics with Applications*, 62(8):3292–3303, 2011.

[133] Khosrow Maleknejad, M. Nosrati, and Esmaeil Najafi. Wavelet Galerkin method for solving singular integral equations. *Computational & Applied Mathematics*, 31(2):373–390, 2012.

[134] Khosrow Maleknejad and M. Nosrati Sahlan. The method of moments for solution of second kind Fredholm integral equations based on B-spline wavelets. *International Journal of Computer Mathematics*, 87(7):1602–1616, 2010.

[135] Khosrow Maleknejad, Monireh Nosrati Sahlan, and Azadeh Ostadi. Numerical solution of fractional integro-differential equation by using cubic B-spline wavelets. In *Proceedings of the World Congress on Engineering*, volume 1, 2013.

[136] Khosrow Maleknejad, M. Shahrezaee, and H. Khatami. Numerical solution of integral equations system of the second kind by block–pulse functions. *Applied Mathematics and Computation*, 166(1):15–24, 2005.

[137] Khosrow Maleknejad and Saeed Sohrabi. Numerical solution of Fredholm integral equations of the first kind by using Legendre wavelets. *Applied Mathematics and Computation*, 186(1):836–843, 2007.

[138] B. N. Mandal and Subhra Bhattacharya. Numerical solution of some classes of integral equations using Bernstein polynomials. *Applied Mathematics and Computation*, 190(2):1707–1716, 2007.

[139] Sergey E. Mikhailov. Analysis of united boundary-domain integro-differential and integral equations for a mixed BVP with variable coefficient. *Mathematical Methods in the Applied Sciences*, 29(6):715–739, 2006.

[140] Kenneth S. Miller and Bertram Ross. *An Introduction to the Fractional Calculus and Fractional Differential Equations.* Wiley, New York, 1993.

[141] R. C. Mittal and Ruchi Nigam. Solution of fractional integro-differential equations by Adomian decomposition method. *The International Journal of Applied Mathematics and Mechanics*, 4(2):87–94, 2008.

[142] M. A. Mohamed and M. Sh Torky. Legendre wavelet for solving linear system of Fredholm and Volterra integral equations. *Int. J. Res. Eng. Sci*, 1:14–22, 2013.

[143] Adel Mohsen and Mohamed El-Gamel. A Sinc–Collocation method for the linear Fredholm integro-differential equations. *Zeitschrift für angewandte Mathematik und Physik*, 58(3):380–390, 2007.

[144] A. Molabahrami, A. Shidfar, and A. Ghyasi. An analytical method for solving linear Fredholm fuzzy integral equations of the second kind. *Computers & Mathematics with Applications*, 61(9):2754–2761, 2011.

[145] Maryam Mosleh and Mahmood Otadi. Fuzzy Fredholm integro-differential equations with artificial neural networks. *Communications in Numerical Analysis*, 2012, Article ID: Cna–00128, 2012.

[146] D. Nazari Susahab and M. Jahanshahi. Numerical solution of nonlinear fractional Volterra-Fredholm integro-differential equations with mixed boundary conditions. *International Journal of Industrial Mathematics*, 7(1):63–69, 2015.

[147] Beny Neta and Paul Nelson. An adaptive method for the numerical solution of Fredholm integral equations of the second kind. I. Regular kernels. *Applied Mathematics and Computation*, 21(2):171–184, 1987.

[148] Robert D. Nevels, Jaideva C. Goswami, and Hooman Tehrani. Semi-orthogonal versus orthogonal wavelet basis sets for solving integral equations. *IEEE Transactions on Antennas and Propagation*, 45(9):1332–1339, 1997.

[149] Van Hoa Ngo. Fuzzy fractional functional integral and differential equations. *Fuzzy Sets and Systems*, 280:58–90, 2015.

[150] Yadollah Ordokhani and Sara Davaei Far. Application of the Bernstein polynomials for solving the nonlinear Fredholm integro-differential equations. *Journal of Applied Mathematics and Bioinformatics*, 1(2):13–31, 2011.

[151] Lu Pan, Xiaoming He, and Tao Lü. High accuracy combination method for solving the systems of nonlinear Volterra integral and integro-differential equations with weakly singular kernels of the second kind. *Mathematical Problems in Engineering*, 2010, Article ID: 901587, 2010.

[152] K. Parand, Mehdi Dehghan, and A. Pirkhedri. The Sinc-collocation method for solving the Thomas–Fermi equation. *Journal of Computational and Applied Mathematics*, 237(1):244–252, 2013.

[153] K. Parand, M. Shahini, and Mehdi Dehghan. Rational Legendre pseudospectral approach for solving nonlinear differential equations of Lane–Emden type. *Journal of Computational Physics*, 228(23):8830–8840, 2009.

[154] Jong Yeoul Park and Jae Ug Jeong. On existence and uniqueness of solutions of fuzzy integro–differential equations. *Indian Journal of Pure and Applied Mathematics*, 34(10):1503–1512, 2003.

[155] H. Parsian. Two dimension Legendre wavelets and operational matrices of integration. *Acta Mathematica Academiae Paedagogicae Nyíregyháziensis*, 21(1):101–106, 2005.

[156] Igor Podlubny. The Laplace transform method for linear differential equations of the fractional order. *arXiv preprint funct-an/9710005*, 1997.

[157] Igor Podlubny. *Fractional Differential Equations: an Introduction to Fractional Derivatives, Fractional Differential Equations, to Methods of their Solution and Some of their Applications*, volume 198. Academic Press, San Diego, 1998.

[158] A. B. Poore. A tubular chemical reactor model. *A collection of nonlinear model problems contributed to the proceedings of the AMS-SIAM*, pages 28–31, 1989.

[159] J. Rashidinia, A. Barati, and M. Nabati. Application of Sinc-Galerkin method to singularly perturbed parabolic convection-diffusion problems. *Numerical Algorithms*, 66(3):643–662, 2014.

[160] J. Rashidinia and M. Zarebnia. New approach for numerical solution of Hammerstein integral equations. *Applied Mathematics and Computation*, 185(1):147–154, 2007.

[161] M. Razzaghi, Y. Ordokhani, and N. Haddadi. Direct method for variational problems by using hybrid of block-pulse and Bernoulli polynomials. *Rom. J. Inf. Sci. Technol*, 2:1–17, 2012.

[162] Owen Willans Richardson. *The Emission of Electricity from Hot Bodies.* Longmans, Green and Company, London, 1921.

[163] Thoudam Roshan. A Petrov–Galerkin method for solving the generalized equal width (GEW) equation. *Journal of Computational and Applied Mathematics*, 235(6):1641–1652, 2011.

[164] Thoudam Roshan. A Petrov–Galerkin method for solving the generalized regularized long wave (GRLW) equation. *Computers & Mathematics with Applications*, 63(5):943–956, 2012.

[165] Yuriy A. Rossikhin and Marina V. Shitikova. Applications of fractional calculus to dynamic problems of linear and nonlinear hereditary mechanics of solids. *Applied Mechanics Reviews*, 50(1):15–67, 1997.

[166] Abbas Saadatmandi and Mehdi Dehghan. Numerical solution of the higher-order linear Fredholm integro-differential-difference equation with variable coefficients. *Computers & Mathematics with Applications*, 59(8):2996–3004, 2010.

[167] Abbas Saadatmandi, Mehdi Dehghan, and Mohammad-Reza Azizi. The Sinc–Legendre collocation method for a class of fractional convection–diffusion equations with variable coefficients. *Communications in Nonlinear Science and Numerical Simulation*, 17(11):4125–4136, 2012.

[168] Abbas Saadatmandi, Mohsen Razzaghi, and Mehdi Dehghan. Sinc-Galerkin solution for nonlinear two-point boundary value problems with applications to chemical reactor theory. *Mathematical and Computer Modelling*, 42(11):1237–1244, 2005.

[169] Jafar Saberi-Nadjafi and Mohamadreza Tamamgar. The variational iteration method: a highly promising method for solving the system of integro-differential equations. *Computers & Mathematics with Applications*, 56(2):346–351, 2008.

[170] S. M. Sadatrasoul and R. Ezzati. Iterative method for numerical solution of two-dimensional nonlinear fuzzy integral equations. *Fuzzy Sets and Systems*, 280:91–106, 2015.

[171] Sh. Sadigh Behzadi and A. Yildirim. A method to estimate the solution of a singular non-linear integro-differential equations by applying the homotopy methods. *International Journal of Industrial Mathematics*, 4(1):41–51, 2012.

[172] S. Saha Ray. On Haar wavelet operational matrix of general order and its application for the numerical solution of fractional Bagley Torvik equation. *Applied Mathematics and Computation*, 218(9):5239–5248, 2012.

[173] S. Saha Ray. A novel method for travelling wave solutions of fractional Whitham–Broer–Kaup, fractional modified Boussinesq and fractional approximate long wave equations in shallow water. *Mathematical Methods in the Applied Sciences*, 38(7):1352–1368, 2015.

[174] S. Saha Ray and R. K. Bera. Analytical solution of the Bagley Torvik equation by Adomian decomposition method. *Applied Mathematics and Computation*, 168(1):398–410, 2005.

[175] S. Saha Ray and A. Patra. Haar wavelet operational methods for the numerical solutions of fractional order nonlinear oscillatory Van der Pol system. *Applied Mathematics and Computation*, 220:659–667, 2013.

[176] S. Saha Ray and P. K. Sahu. Numerical methods for solving Fredholm integral equations of second kind. *Abstract and Applied Analysis*, 2013, Article ID: 426916, 2013.

[177] Santanu Saha Ray. Analytical solution for the space fractional diffusion equation by two-step Adomian decomposition method. *Communications in Nonlinear Science and Numerical Simulation*, 14(4):1295–1306, 2009.

[178] Santanu Saha Ray and A. Patra. Application of homotopy analysis method and Adomian decomposition method for the solution of neutron diffusion equation in the hemisphere and cylindrical reactors. *Journal of Nuclear Engineering & Technology*, 1(2-3):1–12, 2012.

[179] P. K. Sahu and S. Saha Ray. Numerical approximate solutions of nonlinear Fredholm integral equations of second kind using B-spline wavelets and variational iteration method. *Comput. Model. Eng. Sci.*, 93(2):91–112, 2013.

[180] P. K. Sahu and S. Saha Ray. A new approach based on semi-orthogonal B-spline wavelets for the numerical solutions of the system of nonlinear Fredholm integral equations of second kind. *Computational and Applied Mathematics*, 33(3):859–872, 2014.

[181] P. K. Sahu and S. Saha Ray. Numerical solutions for the system of Fredholm integral equations of second kind by a new approach involving semiorthogonal B-spline wavelet collocation method. *Applied Mathematics and Computation*, 234:368–379, 2014.

[182] P. K. Sahu and S. Saha Ray. Legendre wavelets operational method for the numerical solutions of nonlinear Volterra integro-differential equations system. *Applied Mathematics and Computation*, 256:715–723, 2015.

[183] P. K. Sahu and S. Saha Ray. A new numerical approach for the solution of nonlinear Fredholm integral equations system of second kind by using Bernstein collocation method. *Mathematical Methods in the Applied Sciences*, 38(2):274–280, 2015.

[184] Prakash Kumar Sahu and Santanu Saha Ray. Numerical solutions for Volterra integro-differential forms of Lane-Emden equations of first and second kind using Legendre multi-wavelets. *Electronic Journal of Differential Equations*, 2015(28):1–11, 2015.

[185] Prakash Kumar Sahu and Santanu Saha Ray. Two-dimensional Legendre wavelet method for the numerical solutions of fuzzy integro-differential equations. *Journal of Intelligent & Fuzzy Systems*, 28(3):1271–1279, 2015.

[186] P. Salehi and M. Nejatiyan. Numerical method for non-linear fuzzy Volterra integral equations of the second kind. *International Journal of Industrial Mathematics*, 3(3):169–179, 2011.

[187] O. R. Navid Samadi and Emran Tohidi. The spectral method for solving systems of Volterra integral equations. *Journal of Applied Mathematics and Computing*, 40(1-2):477–497, 2012.

[188] S. G. Samko, A. A. Kilbas, and O. I. Marichev. *Fractional Integrals and Derivatives; Theory and Applications.* Gordon and Breach. London, 1993.

[189] K. Sayevand, M. Fardi, E. Moradi, and F. Hemati Boroujeni. Convergence analysis of homotopy perturbation method for Volterra integro-differential equations of fractional order. *Alexandria Engineering Journal*, 52(4):807–812, 2013.

[190] Aydin Secer, Sertan Alkan, Mehmet Ali Akinlar, and Mustafa Bayram. Sinc-Galerkin method for approximate solutions of fractional order boundary value problems. *Boundary Value Problems*, 2013(1):1–14, 2013.

[191] Seppo Seikkala. On the fuzzy initial value problem. *Fuzzy Sets and Systems*, 24(3):319–330, 1987.

[192] A. Shahsavaran. Haar wavelet method to solve Volterra integral equations with weakly singular kernel by collocation method. *Applied Mathematical Sciences*, 5(65):3201–3210, 2011.

[193] Fatemeh Shakeri and Mehdi Dehghan. Solution of a model describing biological species living together using the variational iteration method. *Mathematical and Computer Modelling*, 48(5):685–699, 2008.

[194] W. Smit and H. De Vries. Rheological models containing fractional derivatives. *Rheologica Acta*, 9(4):525–534, 1970.

[195] Ralph C. Smith, Gary A. Bogar, Kenneth L. Bowers, and John Lund. The Sinc-Galerkin method for fourth-order differential equations. *SIAM Journal on Numerical Analysis*, 28(3):760–788, 1991.

[196] Saeed Sohrabi. Comparison Chebyshev wavelets method with BPFs method for solving Abel's integral equation. *Ain Shams Engineering Journal*, 2(3):249–254, 2011.

[197] Shambhunath Srivastava. On the physical validity of Lane-Emden equation of index 5. *Math. Stud*, 34:19–26, 1966.

[198] F. Z. Sun, M. Gao, S. H. Lei, Y. B. Zhao, K. Wang, Y. T. Shi, and N. H. Wang. The fractal dimension of the fractal model of dropwise condensation and its experimental study. *International Journal of Nonlinear Sciences and Numerical Simulation*, 8(2):211–222, 2007.

[199] N. H. Sweilam, M. M. Khader, and W. Y. Kota. On the numerical solution of nonlinear Hammerstein integral equations using Legendre approximation. *International Journal of Applied Mathematical Research*, 1(1):65–76, 2012.

[200] Tao Tang, Xiang Xu, and Jin Cheng. On spectral methods for Volterra integral equations and the convergence analysis. *J. Comput. Math*, 26(6):825–837, 2008.

[201] S. G. Venkatesh, S. K. Ayyaswamy, S. Raja Balachandar, and K. Kannan. Wavelet solution for class of nonlinear integro-differential equations. *Indian Journal of Science and Technology*, 6(6):4670–4677, 2013.

[202] Elder J. Villamizar-Roa, Vladimir Angulo-Castillo, and Yurilev Chalco-Cano. Existence of solutions to fuzzy differential equations with generalized Hukuhara derivative via contractive-like mapping principles. *Fuzzy Sets and Systems*, 265:24–38, 2015.

[203] Hao Wang, H. M. Fu, H. F. Zhang, and Z. Q. Hu. A practical thermodynamic method to calculate the best glass-forming composition for bulk metallic glasses. *International Journal of Nonlinear Sciences and Numerical Simulation*, 8(2):171–178, 2007.

[204] Xing Tao Wang and Yuan Min Li. Numerical solutions of integro–differential systems by hybrid of general block-pulse functions and the second Chebyshev polynomials. *Applied Mathematics and Computation*, 209(2):266–272, 2009.

[205] Abdul-Majid Wazwaz. A new algorithm for solving differential equations of Lane–Emden type. *Applied Mathematics and Computation*, 118(2):287–310, 2001.

[206] Abdul-Majid Wazwaz. A new method for solving singular initial value problems in the second-order ordinary differential equations. *Applied Mathematics and computation*, 128(1):45–57, 2002.

[207] Abdul-Majid Wazwaz. Analytical solution for the time-dependent Emden–Fowler type of equations by Adomian decomposition method. *Applied Mathematics and Computation*, 166(3):638–651, 2005.

[208] Abdul-Majid Wazwaz. *Linear and Nonlinear Integral Equations: Methods and Applications.* Springer Science & Business Media, Berlin, 2011.

[209] Abdul-Majid Wazwaz. The variational iteration method for solving the Volterra integro-differential forms of the Lane–Emden equations of the first and the second kind. *Journal of Mathematical Chemistry*, 52(2):613–626, 2014.

[210] Abdul-Majid Wazwaz, Randolph Rach, and Jun-Sheng Duan. Adomian decomposition method for solving the Volterra integral form of the Lane–Emden equations with initial values and boundary conditions. *Applied Mathematics and Computation*, 219(10):5004–5019, 2013.

[211] Congxin Wu and Zengtai Gong. On Henstock integral of fuzzy-number-valued functions (I). *Fuzzy Sets and Systems*, 120(3):523–532, 2001.

[212] Lan Xu, Ji-Huan He, and Yong Liu. Electrospun nanoporous spheres with Chinese drug. *International Journal of Nonlinear Sciences and Numerical Simulation*, 8(2):199–202, 2007.

[213] Salih Yalçınbaş and Tuğçe Akkaya. A numerical approach for solving linear integro-differential-difference equations with Boubakcr polynomial bases. *Ain Shams Engineering Journal*, 3(2):153–161, 2012.

[214] LiJun Yi and BenQi Guo. An hp Petrov-Galerkin finite element method for linear Volterra integro-differential equations. *Science China Mathematics*, 57(11):2285–2300, 2014.

[215] Mingxu Yi and Jun Huang. CAS wavelet method for solving the fractional integro-differential equation with a weakly singular kernel. *International Journal of Computer Mathematics*, 92(8):1715–1728, 2015.

[216] Ahmet Yıldırım and Turgut Öziş. Solutions of singular IVPs of Lane–Emden type by homotopy perturbation method. *Physics Letters A*, 369(1):70–76, 2007.

[217] S. Yousefi and M. Razzaghi. Legendre wavelets method for the nonlinear Volterra–Fredholm integral equations. *Mathematics and Computers in Simulation*, 70(1):1–8, 2005.

[218] S. A. Yousefi. Legendre multiwavelet Galerkin method for solving the hyperbolic telegraph equation. *Numerical Methods for Partial Differential Equations*, 26(3):535–543, 2010.

[219] S. A. Yousefi, Zahra Barikbin, and Mehdi Dehghan. Ritz-Galerkin method with Bernstein polynomial basis for finding the product solution form of heat equation with non-classic boundary conditions. *International Journal of Numerical Methods for Heat & Fluid Flow*, 22(1):39–48, 2012.

[220] S. A. Yousefi and M. Behroozifar. Operational matrices of Bernstein polynomials and their applications. *International Journal of Systems Science*, 41(6):709–716, 2010.

[221] S. A. Yousefi, M. Behroozifar, and Mehdi Dehghan. Numerical solution of the nonlinear age-structured population models by using the operational matrices of Bernstein polynomials. *Applied Mathematical Modelling*, 36(3):945–963, 2012.

[222] Sohrab Ali Yousefi. Legendre wavelets method for solving differential equations of Lane–Emden type. *Applied Mathematics and Computation*, 181(2):1417–1422, 2006.

[223] ElçIn YusufoğLu. A homotopy perturbation algorithm to solve a system of Fredholm–Volterra type integral equations. *Mathematical and Computer Modelling*, 47(11):1099–1107, 2008.

[224] Şuayip Yüzbaşı, Mehmet Sezer, and Bayram Kemancı. Numerical solutions of integro-differential equations and application of a population model with an improved Legendre method. *Applied Mathematical Modelling*, 37(4):2086–2101, 2013.

[225] Lotfi A. Zadeh. The concept of a linguistic variable and its application to approximate reasoning–I. *Information Sciences*, 8(3):199–249, 1975.

[226] M. Zarebnia. Solving nonlinear integral equations of the Hammerstein-type by using double exponential transformation. *Australian Journal of Basic and Applied Sciences*, 4(8):3433–3440, 2010.

[227] Mohsen Zayernouri, Mark Ainsworth, and George Em Karniadakis. A unified Petrov–Galerkin spectral method for fractional PDEs. *Computer Methods in Applied Mechanics and Engineering*, 283:1545–1569, 2015.

[228] Jingjun Zhao, Jingyu Xiao, and Neville J Ford. Collocation methods for fractional integro-differential equations with weakly singular kernels. *Numerical Algorithms*, 65(4):723–743, 2014.

[229] Xiaoyang Zheng and Xiaofan Yang. Techniques for solving integral and differential equations by Legendre wavelets. *International Journal of Systems Science*, 40(11):1127–1137, 2009.

[230] Li Zhu and Qibin Fan. Solving fractional nonlinear Fredholm integro-differential equations by the second kind Chebyshev wavelet. *Communications in Nonlinear Science and Numerical Simulation*, 17(6):2333–2341, 2012.

[231] Li Zhu and Qibin Fan. Numerical solution of nonlinear fractional-order Volterra integro-differential equations by SCW. *Communications in Nonlinear Science and Numerical Simulation*, 18(5):1203–1213, 2013.

[232] Hua Zou and Hua Li. A novel meshless method for solving the second kind of Fredholm integral equations. *Computer Modeling in Engineering and Sciences (CMES)*, 67(1):55, 2010.

Index

Atlantic
26-3-21

£ 115-00